Victor Hugo Gomes Sales
Paulo Victor Gomes Sales
Joenes Mucci Peluzio

Multivariate analysis of soybeans in different parts of the plant

Victor Hugo Gomes Sales
Paulo Victor Gomes Sales
Joenes Mucci Peluzio

Multivariate analysis of soybeans in different parts of the plant

Aiming at energy production

ScienciaScripts

Imprint
Any brand names and product names mentioned in this book are subject to trademark, brand or patent protection and are trademarks or registered trademarks of their respective holders. The use of brand names, product names, common names, trade names, product descriptions etc. even without a particular marking in this work is in no way to be construed to mean that such names may be regarded as unrestricted in respect of trademark and brand protection legislation and could thus be used by anyone.

Cover image: www.ingimage.com

This book is a translation from the original published under ISBN 978-620-2-03218-6.

Publisher:
Sciencia Scripts
is a trademark of
Dodo Books Indian Ocean Ltd. and OmniScriptum S.R.L publishing group

120 High Road, East Finchley, London, N2 9ED, United Kingdom
Str. Armeneasca 28/1, office 1, Chisinau MD-2012, Republic of Moldova, Europe
Managing Directors: Ieva Konstantinova, Victoria Ursu
info@omniscriptum.com

Printed at: see last page
ISBN: 978-620-6-75946-1

DEDICATORY

To my parents, Celso dos Reis Sales in memorian and Mara Rúbia Gomes Sales, for being role models in my life. I also dedicate it to my wife Elisa Maria de Oliveira and my daughters, for their understanding, strength, help and companionship that make my life complete.

ACKNOWLEDGEMENTS

Firstly, I would like to thank God for my life and for all the love, mercy, consolation, protection, strength, education and guidance he has given me.

I would like to thank my advisor, Professor Dr Joenes Mucci Peluzio, for helping me to carry out this work and for his dedication, teachings and friendship throughout my master's degree.

To the professor and coordinator of the postgraduate programme, Dr Emerson Adriano Guarda, for his teachings, advice, support and friendship during my master's degree.

The Federal University of Tocantins - Palmas Campus, especially the Postgraduate Programme in Agroenergy.

CAPES for the grant awarded during my master's degree, without which it would not have been possible to carry out this work.

To my wife, Elisa Maria de Oliveira, for her affection, strength, companionship and understanding during this period, to our daughters, Maria Clara de Oliveira Gomes Sales and Mariana de Oliveira Gomes Sales for their understanding and affection, they are the reason for my existence.

To my mum Mara Rúbia, my brothers Michelle and Paulo Victor, my sisters-in-law Ana Claudia and Isabel, my brother-in-law Eder Terra and my nieces and nephews Maria Fernanda, Maria Eduarda, Amanda Vitória, Celso Neto, João Guilherme, Amanda and Victor Hugo, for all their affection, love, effort and support. I love you all!

To all the teachers, colleagues and staff at UFT who helped me along this journey. Those who made themselves available at any time, without naming names, so as not to risk forgetting anyone, because there were so many who helped make this happen.

/ Revista Agro@mbiente On-line of the Centre for Agricultural Sciences of the UFRR for the prior authorisation to publish the second chapter of the book.

SUMMARY

INTRODUCTION

Soya *[Glycine max* (L.) Merril], a plant native to China, is considered the most important oilseed grown in the world (EDMAR et al, 2008; CLEMENTE & CAHOON, 2009). The crop's commercial interest is related to its high protein and oil content, on average 40 per cent and 20 per cent respectively, as well as its productivity (BONATO et al., 2000; SEDIYAMA et al., 2005 and VELLO & SILVA, 2006). In addition, the grain contains around 34% carbohydrates and 5% minerals (EMBRAPA, 2013).

As well as being an important food source of protein and oil, soya is also rich in minerals such as iron, calcium, phosphorus, potassium and B vitamins (GAVIOLI et al, 2012). It contains bioactive compounds, such as isoflavones, which have been widely studied due to their beneficial effects on human health in controlling cancer, cardiovascular problems, osteoporosis, cholesterol, diabetes, obesity and climacteric syndrome (MANDARINO & CARRÃO-PANIZZI, 1997; LUI et al., 2003; GAVIOLI, 2012).

As soya oil has good digestibility and no cholesterol, it accounts for 92% of edible oil consumption (SEDYAMA, 2009). The oil has a high content of polyunsaturated acids, such as linoleic acid and a-linoleic acid, with 55% and 8% respectively (DIXIT, 2011). These characteristics maximise the possibility of extracting the oil for food, animal feed and biofuel production, as well as producing bran with high protein content for food (AMORIM, 2011).

National biofuel programmes have been stimulating the supply of vegetable oils in the biofuel market, and soybeans stand out in this scenario because they are widely adapted to the climate (North-South), have a broad technological domain (oil and protein - multipurpose), favourable logistics and resistance to market price fluctuations. These factors have a positive impact on the high production and supply of the crop (FERRARI et al., 2005).

Another important application of the soya bean crop in the field of Agroenergy is the use of the crop not only as a raw material for the production of biodiesel, but for a range of other functions, ranging from the generation of electricity from the burning of solid waste in boilers (GOBIRA, 2012). The production of ethanol from the crop has been gaining ground in Brazil. Ethanol can be produced from the fermentation of soya molasses, which is a by-product obtained from the production of concentrated bran (VELLO & SILVA, 2006).

The protein and oil content of soya beans can vary depending on a number of factors, including: climatic conditions (BARBOSA et al., 2011); temperature (VOLLMANN et al. 2000); cultivars (SOUZA et al., 2009); fertiliser levels (TOLEDO et al., 2010); leaf and pod removal (PROULX & NAEVE, 2009); shading (PROULX & NAEVE, 2009), 2009); fertiliser levels (TOLEDO et aL, 2010); leaf and pod removal (PROULX & NAEVE, 2009); shading (PROULX & NAEVE, 2009; BELLALOUI & GILLEN, 2010) and the position of the pods at the nodes of the main stem (BENNETT et aL, 2003; BELLALOUI & GILLEN, 2010; SALES et al. 2013).

In relation to the position of the pods on the nodes of the main stem of the plant, Benett et al. (2003), in soya cultivation under field conditions and Bellaloui & Gillen (2010), in cultivation under greenhouse conditions, both studies carried out in the United States and Sales et al. (2013), in trials carried out under field conditions in Brazil, in the city of Palmas, Tocantins state, observed a tendency for the oil content to be more concentrated in the beans located in the lower third of the plants and the protein content to be more concentrated in the upper third. On the other hand, in the research carried out by Huskey et al. (1990), no significant difference was observed in the protein content of soya beans between the upper and lower thirds of the plant, which in turn were higher than the middle third.

Some agronomic characteristics such as the number of pods can vary according to certain factors, such as: number of flowers per node on the plant (JIANG & EGLI, 1993); luminosity (EGLI et al.,1987); shading due to limitation of photoassimilates (BELLALOU & GILLEN, 2010), which will negatively influence seed size and consequently unit weight, hundred-seed weight and yield (BOARD et al.,1997; MIRANDA, 1998).

Average seed weight is a genetically determined characteristic that is affected by environmental factors (PANDEY & TORRIE, 1973). Seed size has a positive and negative influence on some production factors, such as plant height, number of pods, number of seeds and yield (EDWARD JÚNIOR & HARTWING, 1971; GONTIA & AWASTHI, 1999; KRZYZANOWSKI, 2004; 2005; 2008; COSTA, 2006). There is a lot of controversy regarding the effect of soya bean seed size on the crop's performance in the field, and more research is needed (LIMA & CARMONA, 1999)

The effect of shading and the consequent limitation of photoassimilates due to the incidence of low illumination in the lower third will lead to low photosynthetic assimilation in the lower leaves (BELLALOU & GILLEN, 2010), damaging pod formation and, as a result, physically limiting seed size, average weight and production (EGLI et al. 1987).

In view of the above and the scarcity of studies on the effect of the position of the pods on the main stem of the plant on protein and oil content and some agronomic characteristics under low latitude conditions, this work was carried out with the aim of: a) studying the effect of the position of the pods on oil and protein content in different parts of the plant; b) the genetic diversity of soya cultivars in different parts of the plant, through multivariate analysis using the TOCHER clustering method, obtained from the Mahalanobis distance.

CHAPTER 1

LITERATURE REVIEW

Origin and distribution of soya

The soya bean was originally a type of creeping plant that grew along rivers and lakes, very different from the one cultivated today. Its evolution began with the appearance of plants originating from natural crosses between two species of wild soya that were domesticated and improved by scientists in ancient China (REVISTA FUNCIONAIS & NUTRACÊUTICOS, 2007).

Hymowitz (1970) and Bonato & Bonato (1987) report that the crop was domesticated in the northern half of China around the 11th century BC. According to Hymowitz (1990), domestication is a process of trial and error rather than an event, and in the case of soya, this process probably took place during the Shang dynasty (1700-1100 BC). By the first century after Christ, soya probably reached central and southern China, as well as the Korean peninsula.

In Brazil, the crop was introduced in Bahia in 1882 by Gustavo D'utra (BONATO & BONATO, 1987) and cultivated for the first time at the Campinas Agricultural Station in 1901. The grain then arrived with greater intensity with the first Japanese immigrants in 1908 and was officially introduced in Rio Grande do Sul in 1914. However, the expansion of soya in Brazil took place in the 1970s with the growing interest of the oil industry (REVISTA FUNCIONAIS & NUTRACÊUTICOS, 2007).

Agricultural production in the state of Tocantins has advanced significantly since its creation. In 1990, the planted area was 344.58 thousand hectares, compared to 730.97 thousand hectares in 2012, representing an increase of approximately 112.13%. Over the last 25 years, the crop has made significant progress, with an increase of 1940% in the total area planted, 3525% in grain production and 181% in productivity. This increase can be justified initially by advances in the genetic improvement of the crop and the discovery of varieties adapted to the state's soil and climate conditions.

According to the IBGE Automatic Data Retrieval System (SIDRA) in 2012, the main agricultural crops, with their respective planted areas and main producing municipalities in the state were: Pineapple (1,699 ha - Miracema do Tocantins); Cotton (7,500 ha - Porto Nacional); Peanuts (2,500 ha - Tupiratins); Rice (111,155 ha - Lagoa da Confusão and Formoso do Araguaia); Sugarcane (27,405 ha - Pedro Afonso); Beans (30,321 ha - Lagoa da Confusão); Cassava (16.535 ha - Porto Nacional); Watermelon (4,036 ha - Lagoa da Confusão and Formoso do Araguaia); Maize (96,973 ha - Campos Lindos); Soya (415,463 ha - Campos Lindos, Mateiros, Dianópolis, Porto Nacional and Lagoa da Confusão) and Sorghum (17,320 ha - Campos Lindos and Pedro Afonso) (IBGE, 2013).

Economic importance

According to Edmar et al. (2008) and Clemente & Cahoon (2009), soya is the most important oilseed crop grown in the world. However, it was only at the beginning of the last century that it took centre stage in the Western world (ZANON et al, 2007; GOMES et al, 1988; BONETT et al, 1999).

According to Silva et al (2006) and Espíndola et al (2008), in Brazil soya is considered a *commodity* of great commercial importance to the Brazilian economy, occupying large tracts of land planted in the cerrado region. Silva et al. (2010) reports that the increase in demand for soya production is mainly due to the rise in global demand for proteins for animal feed, resulting in the expansion of international trade in soya and its products, changing the country's economic policy and benefiting the production chain.

According to Bonato et al. (2000), Sedyama et al. (2005) and Vello & Silva (2006), the commercial interest in soya is related to its high protein and oil content, averaging 40% and 20% respectively, and also to its productivity. Another important characteristic of soya, according to Amorim (2011), is the crop's ability to fix nitrogen symbiotically in its roots, thus eliminating the need for nitrogen fertilisers.

Chemical composition of grains

According to Machado (1999), the protein fraction of soya is made up of a large number of nitrogenous compounds, not only in terms of quantity but also in terms of the quality of its protein. According to Dixit et al. (2011), the majority of soya proteins are classified as globulins, around 90%, and the globulins present in soya are glycine (11S fraction) and p-conglycine (7S fraction).

Soya proteins contain all the essential amino acids for human nutrition, which makes soya products almost equivalent to animal sources in terms of protein quality, but with less saturated fat and no cholesterol. Soya also contains the biologically active protein components of haemagglutinin, trypsin inhibitors, a-amylase and lipoxygenases (DIXIT et al., 2011).

Both fatty acids are considered essential, from the w-6 and (JÚ-3 family, respectively, which have important physiological and nutritional functions and regulate a series of metabolic pathways. Herbinet et al. (2008) found that the grain also has a high oleic acid content, approximately 30 per cent.

Another important component is phosphatides, which are compounds similar to glycerides, but made up of fatty acids linked to nitrogen, or nitrogen and phosphorus. The main compounds in this group are lecithins, cephalins and sphingomyelins. In soya beans, these components are around 1.8% phosphatides and can reach 3.2% (MACHADO, 1999).

Unsaponifiable matter represents around 0.52 to 1.14 per cent of the total lipids in soya, which includes a wide range of compounds such as pigments (carotenoids, isoflavonoids, anthocyanins and chlorophyll), sterols (ergosterol and sitoesterol), tocopherols, tocokinones, hydrocarbons, alcohols, ketones, aldehydes, vitamins and others (MACHADO, 1999).

Soya contains approximately 35% carbohydrates, most of which are polysaccharides, oligosaccharides such as stachyose and raffinose with 4%

and 1.1% concentration in the grain, respectively (DIXIT, 2011). Soya beans also contain cellulose with approximately 64%, which is mainly present in the husk (MACHADO, 1999).

Soya carbohydrates are little explored as a food source, being used mainly in animal feed (mainly for ruminants) as a source of calories. It should be noted that although soya sugars are complex, animal feed does not undergo any processing, as there is no added nutritional value from the breakdown of sugars in the feed (IRISH, 1998).

The economic value of soya carbohydrates is considered small compared to the protein and oil content. However, research has recently been carried out into how soya carbohydrates can be used to produce ethanol from soya molasses, which is made by reusing the industrial waste obtained from the process of producing soya protein concentrates.

According to Machado (1999), in addition to the savings generated by self-sufficiency in solvent production, ethanol from soya molasses is more advantageous than ethanol from sugar cane because it has reduced some costs such as transporting the cane, milling, filtering the juice and handling the bagasse.

According to Dixit (2011), soya is the best source of B vitamins among cereals, with the exception of vitamin B12 and vitamin C. It also contains approximately 5% of minerals, including K, P, Ca, Mg, Fe and others. Soya also contains ferritin, an iron storage protein, which is much more absorbed and bioavailable than animal products.

According to Machado (1999) the lipid fraction of soya beans is made up mainly of glycerides, phosphatides and a small portion of non-saponifiable matter, called spherosomes, where glycerides, which are the main components of the lipid fraction, are made up of a glycerol molecule linked to three fatty acids.

Soya oil accounts for 92% of edible oil consumption, contains no cholesterol and has good digestibility (SEDIYAMA, 2009). According to Dixit

(2011) the oil has a high content of polyunsaturated fatty acids, namely linoleic acid and a-linoleic acid with 55% and 8% respectively.

These characteristics maximise the possibility of extracting oil for both human consumption and biodiesel production, as well as producing a bran with high protein content for animal feed, favouring the socio-economic value of the crop (AMORIM, 2011).

Factors that influence the chemical composition of soya beans

The chemical composition of soya is considerably influenced by the genetics of the varieties (MACHADO, 1999; ÁVILA et al., 2007; SOUZA et al., 2009) and the climatic conditions in which the plants grow (MACHADO, 1999, ALBRECHT et al., 2008; BARBOSA et al., 2011), especially the amount of light (MACHADO, 1999).

Other factors that influence the protein and oil composition of soya beans have been observed for fertiliser levels (MORAES *et al.,* 2008; Toledo *et al.*, 2010, VEIGA *et al.*, 2010), shading (PROULX & NAEVE, 2009; BELLALOUI & GILLEN, 2010), the position of the pods at the nodes of the main stem (COLLINS & CARTTER, 1956; HUSKEY et al.,1990ESCALANTE & WILCOX, 1993; BENETT et al, 2003; BELLALOUI & GILLEN, 2010; SALES et aL, 2013), pod removal (ROCHA et al. 1996; PELUZIO et al. 2001; PROULX & NAEVE, 2009) and leaf removal (CAMPELO et al. 1999; PRATISSOLI et al. 2001; PELUZIO et al. 2002; PELUZIO et al. 2004; PROULX & NAEVE, 2009).

In a trial carried out in the 2010/2011 agricultural year with 10 transgenic soya cultivars in the city of Palmas, Tocantins state, SALES et aL, (2013), observed differences in the percentage of oil in the beans, depending on the location of the pods on the main stem of the plant, with the highest concentration of oil in the lower third of the plant, followed by the middle and upper thirds. The opposite was observed for protein content, where the highest values were concentrated in the upper third, followed by the middle and lower thirds (SALES, 2012).

Biodiesel production and supply chain

With a well-structured production chain, soya cultivation plays an important role in the biodiesel issue, as it has the production and technical capacity to meet part of the growing demand for renewable fuels, given that natural energy resources such as fossil fuels have been consumed at high rates in recent years. What's more, as the rate at which fossil fuels are replaced is lower than their consumption, there may be a reduction in their reserves, raising the final price of the product on the market. In this sense, with the aim of becoming self-sufficient, reducing spending on diesel imports, the Brazilian government created the National Agroenergy Plan 2006-2011, increasing the addition of biodiesel to diesel in different proportions, according to the targets to be achieved (GOBIRA, 2012).

According to Law No. 11,097 of 13 January 2005, biodiesel is defined as a biofuel derived from renewable biomass for use in internal combustion engines with compression ignition or, according to regulations, for generating other types of energy, which can replace partially or totally fossil fuels. It can be used pure (B100) or mixed with diesel in different proportions, where a mixture of 5% biodiesel with petroleum diesel is called B5 and so on. Technically, biodiesel is defined as a mono-alkyl ester of fatty acids derived from renewable sources (MONYEM et al., 2001), which through transesterification transform triglycerides into smaller molecules of fatty acid esters.

According to Durães (2010), biodiesel production faces a number of bottlenecks that constitute major opportunities and risks: technical-scientific challenges in both the agronomic and industrial areas, availability of raw materials, availability of modern inputs for agroenergy, waste and co-products in the production chains, machinery and implements, investment and management, the market and logistics. In the agronomic area, investment in research is the basis for developing agricultural production technologies, identifying species with exploitation potential, more efficient production systems and selecting regions with high biodiesel production potential.

Biodiesel can be obtained by different processes such as esterification and transesterification, and can be produced from animal fats or vegetable oils from soya, sunflower, cotton, castor and palm. The most widely used process for producing biodiesel today is transesterification, which consists of a chemical reaction of vegetable oil or animal fat with ethanol or methanol, stimulated by catalysis, generating glycerine in addition to biodiesel. The process of transesterifying vegetable oils produces esters with a viscosity approximately double that of diesel (KNOTHE et al., 2005).

In 2010, 600 million litres of biodiesel were sold, at least 80% of which came from producers with the Social Fuel Seal, at auctions held by the National Petroleum, Natural Gases and Fuels Agency (ANP). This auctioned volume represents the entire demand for biodiesel to meet the legal 5% blend in diesel in the period from January to March 2011 (MME, 2011).

Sandun et al (2005), in his work, states that even though there are several oilseeds identified with the potential to be used in biodiesel production, only soya, sunflower, rapeseed, cotton, palm and jatropha can meet the technical specifications of the American Standards for Testing of Materials (ASTM D 6751). This is due, among other factors, to viscosity, which is an important property of any fuel as it is an indicator of flow capacity (PETERSON et al, 1999; SANTOS, 2007). The high viscosity of vegetable oils is due to their high molecular weight and large chemical structure. The viscosity of pure vegetable oils is 10 to 15 times higher than the viscosity of diesel oil.

In Brazil, 90 per cent of the vegetable oil produced comes from soya, even though it has a lower oil yield per area compared to other crops, due to the following factors: 1) a well-structured production chain; 2) well-defined production technologies, supported by a broad research network; 3) a fast cycle, offering a quick return on investment (4 to 5 months); 4) the crop is adapted to produce uniformly throughout the country.

Currently, around 85% of biodiesel production comes from soya oil, although other oilseeds have a much higher oil content, these still need research and supply on a large scale.

Another important application of the soya bean crop in the field of Agroenergy is reported by Gobira (2012), which is the use of the crop not only as a raw material for the production of biodiesel, but for a range of other functions, ranging from the generation of electricity, from the burning of solid waste in boilers, to the production of ethanol, which can be produced from soya molasses, which is a by-product obtained from the production of concentrated bran (VELLO & SILVA, 2006).

National biofuel programmes have been stimulating the supply of vegetable oils on the biofuel market, and soya beans, due to their wide climate adaptation (North-South), broad technological domain (oil and protein - multi-purpose), favourable logistics and resistance to market price fluctuations, reflect positively on the high production and supply of the crop.

Genetic divergence and multivariate analysis

For many types of biological data, there is a correlation between variables and the information provided by isolated univariate analyses can be incomplete. Thus, multivariate analysis techniques simultaneously combine multiple pieces of information from an experimental unit that cannot be obtained using univariate analysis (VIANA et aL, 2000).

The use of multivariate techniques to estimate genetic divergence has been used in several studies and in various crops, such as: rice (BENITEZ et aL, 2011); coriander (BERTINI et aL, 2010); cherry tomatoes (ROCHA et aL, 2010); sugar cane (DUTRA FILHO, 2011); castor beans (BEZERRA NETO et aL, 2010); cassava (VENDRAMINI et aL, 2011); aubergine (ARAMENDIZ-TATIS et aL, 2011); maize (COIMBRA et aL, 2010); sunflower (VOGT, BALBINOT JÚNIOR & SOUZA, 2010); melon (NUNES et aL, 2011); eucalyptus (CASTRO et aL, 2013) and soya (PELUZIO et al., 2009; SANTOS et al., 2011; ALMEIDA, PELUZIO & AFFÉRRI, 2011 and SANTOS et al., 2013).

Among multivariate techniques, the most commonly used are grouping methods (CRUZ & REGAZZI, 2004). The use of multivariate techniques is one of the factors driving the increase in studies on genetic divergence between germplasm bank accessions. Multivariate analyses are based on algorithms, or distance measures, which simultaneously take into account numerous characteristics considered in germplasm characterisation and evaluation experiments (SUDRÉ et al., 2007; VILELA et al., 2008; ROCHA, 2010).

The purpose of clustering methods is to separate an original group of observations into several subgroups in order to obtain homogeneity within and heterogeneity between the subgroups. This methodology depends on the calculation of dissimilarity measures from quantitative and qualitative variables (CROSSA & FRANCO, 2004; ROCHA, 2010). Among these methods, hierarchical and optimisation methods are widely used by plant breeders (CRUZ, 2007).

In optimisation methods, groups are established by perfecting a certain grouping criterion, differing from hierarchical methods in that the groups formed are mutually exclusive (CRUZ & REGAZZI, 2004). In the optimisation method proposed by TOCHER, the criterion of keeping the average intra-group distance always lower than any inter-group distance is adopted (RAO, 1952).

BIBLIOGRAPHICAL REFERENCES

ALBRECHT, L.P.; BRACCINI, A.L.; ÁVILA, M.R.; SUZUKI, L.S.; SCAPIM, C.A.; BARBOSA, M.C. Oil and protein content and yield of soya beans as a function of early sowing in western Paraná. **Bragantia,** v. 67, n. 04, p. 865-873, 2008.

ALMEIDA, R. D.; PELUZIO, J. M.; AFFÉRRI, F. S. Genetic divergence between soya cultivars under irrigated floodplain conditions in southern Tocantins State. **Revista Ciência Agronómica,** v.42, n.01, p. 108-115, 2011.

AMORIM, F. A. **Effect of the CP4 EPSPS gene on oil productivity in soya populations.** 2011. Thesis (PhD in Genetics and Plant Breeding) - Luiz de Queiroz College of Agriculture, University of São Paulo, Piracicaba, 2011. Available at:< http://www.teses.usp.br/teses/disponiveisZ11/11137/tde-13092011 - 083447/>. Accessed on: 03 Mar 2013.

ARAMENDIZ-TATIS, H.; SUDRÉ, C. P; GONÇALVES, L. S. A; RODRIGUES, R. Agronomic potential and genetic divergence among aubergine genotypes under Colombian Caribbean conditions. **Horticultura Brasileira,** v.29, p.174-180, 2011.

ÁVILA, M.R.; BRACCINI, A.L.; SCAPIM, C.A.; MANDARINO, J.M.G.; ALBRECHT, L.P.; FILHO, P.S.V. Yield Components, Isoflavone, Protein and Oil Content and Quality of Soya Seeds. **Revista Brasileira de Sementes,** v. 29, n. 3, p.111-127, 2007.

BARBOSA, V.S.; PELUZIO, J.M.; AFFÉRRI, F.S.; SIQUEIRAS, G.B. Behaviour of soya cultivars at different sowing times for biofuel production. **Revista Ciência Agronómica,** v.42, n.3, p.742-749, 2011.

BELLALOUI, N.; GILLEN, A. M.; Soybean seed protein, oil, fatty acids, N, and S partitioning as affected by node position and cultivar differences. **Agricultural Sciences, v. 1, n. 3,** p. 110-118, 2010.

BENITEZ, L.C.; RODRIGUES, I.C.S.; ARGE, W.P.; RIBEIRO, M.V.; BRAGA, E.J.B. Multivariate analysis of the genetic divergence of rice genotypes under salt stress during the vegetative phase. **Revista Ciência Agronómica.** v.42, n.02, p.409-416, 2011.

BENNETT, J.O.; KRISHNAN, A.H.; WIEBOLD, W.J.; KRISHNAN, H.B. Positional Effect on Protein and Oil Content and Composition of Soybeans. **Agricultural And Food Chemistry, v. 51,** p. 6882-6883, 2003.

BERTINI, C.H. M.; PINHEIRO, E.A.R.; NÓBREGA, G.N.; DUARTE, M.L. Agronomic performance and genetic divergence between coriander genotypes. **Revista Ciência Agronómica,** v. 41, n. 03, p. 409-416, 2010.

BEZERRA NETO, F.V.; LEAL, N.R.; GONÇALVES, L.S.A.; REGO FILHO, L.M.; AMARAL JÚNIOR, A.T. Quantitative descriptors in the estimation of genetic divergence between castor bean genotypes using multivariate analyses. **Revista Ciência Agronómica,** v.41, n.02, p.294-299, 2010.

BOARD, J.E.; KANG, M.S.; HARVILLE, B.G. Path analysis identifies indirect selection criteria for yield of late planted soybean. **Crop Science,** v. 37, n. 03, p. 879-884, 1997.

BONATO, E.R.; BONATO, A.L.V. **Soya in Brazil: History and statistics.** Londrina: EMBRAPA-CNPSo, 1987. 61 p.

BONATO, E.R.; BERTAGNOLLI, P.F.; LANGE, C.E.; RUBIN, S. de A.L.Oil and protein content in soya genotypes developed after 1990. **Pesquisa Agropecuária Brasileira,** v. 35, p. 2391-2398, 2000.

BONETT, L.P.; GONÇALVES-VIDIGAL, M.C.; SCHUELTER, A.R.; VIDIGAL FILHO, P.S.; GONELA, A.; LACANALLO, G.F. Genetic divergence in common bean germplasm collected in the State of Paraná, Brazil. **Semina: Ciências Agrárias,** v.27, p.547-560, 2006

CAMPELO, J.E.G.; SEDIYAMA, T.; ROCHA, V.S.; SEDIYAMA, C.S.; Effects of defoliation on the onset of flowering in winter soya beans. **Revista Ceres,** v. 46, n. 268, p. 652-666,1999.

CASTRO, A.F.N. M.; CASTRO, R.V.O.; CARNEIRO, A.C.O.; LIMA, J.E.; SANTOS, R.C.; ALVES, I.C.N. Multivariate analysis for the selection of eucalyptus clones for charcoal production. **Pesquisa Agropecuária Brasileira,** v.48, n.6, p.627-635, 2013.

CLEMENTE, T.E.; CAHOON, E.B. Soybean Oil: Genetic Approaches for Modification of Functionality and Total Content. **Plant Physiology,** v. 151, p. 1030-1040, 2009.

COIMBRA, R.R.; MIRANDA, G.V.; CRUZ, C.D.; ECKERT, F.R. Characterisation and genetic divergence of maize populations rescued from the Southeast of Minas Gerais. **Revista Ciência Agronómica,** v.41, n.01, p.159-166, 2010.

COSTA, P.R. **Water deficiency: Study of osmotic agents and seed size.** Presidente Prudente, 2006. 57 p. Dissertation (Master's Degree in Agronomy) - Universidade Oeste Paulista.

COLLINS, F.I.; CARTTER, J.L.; Variability in Chemical composition of seed from different portions of the soybean plant. **Agronomy Journal,** v. 48, p. 216-219, 1956.

CONAB - NATIONAL SUPPLY COMPANY. Monitoring the Brazilian harvest: Grains - 2013/2014 harvest - First

survey. Brasília: Conab, 2013. 67 p. Available at <http://www.conab.gov.br/OlalaCMS/uploads/arquivos/13 10 23 13 46 38 boletim_portugues_outubro_2013.pdf>. Accessed on 08 Nov. 2013.

GROSSA, J.; FRANCO, J. Statistical methods for classifying genotypes. **Euphytica,** v. 137, n. 1, p. 19-37, 2004.

CRUZ, C.D. **Programa Genes: Aplicativo computacional em genética e estatística.** Windows version - 2007, Viçosa, UFV.

CRUZ, C.D.; REGAZZI, A.J. **Biometric models applied to genetic improvement.** Viçosa, MG: University Press, 2004, 480 p.

DURÃES, F.O.M. **Ponto de vista - Agroenergy for oils and co-products:**

Palms for biodiesel. EMBRAPA AGROENERGIA. p.04. Brasília-DF - 2010 EMBRAPA. Brazilian livestock research company. Chemical composition and properties. On online. Available at: <http://www.cnpso.embrapa.br/soja_alimentacao/index.php?pagina=23>. Accessed on 18 January 2014.

DUTRA FILHO, J.A.; MELO, L.J.O.; RESENDE, L.V.; ANUNCIAÇÃO FILHO, C.J.; BASTOS, G.Q. Application of multivariate techniques in the study of genetic divergence in sugarcane. **Revista Ciência Agronómica,** v.42, n.01, p.185-192, 2011.

EDMAR, V.S.; REIS, M.S.; SEDIYAMA, T. Non-parametric analysis of seed health and elimination indices and classification of soya genotypes. **Pesquisa Agropecuária Brasileira,** v. 43, p. 341-348, 2008.

EDWARD JÚNIOR, C.J.; HARTWING, E.E. Effect of seed size upon rate of germination in soybeans. **Agronomy Journal,** v.4, p.13-30,1980.

EGLI, D.B.; WIRALAGA, R.A.; BUSTAMAM, T.; YU,Z.W.; TEKRONY, D.M. Time of flower opening and seed mass in soybean. **Agronomy Journal,** v.79, n.4, p.697-700, 1987.

EMBRAPA - Brazilian Agricultural Research Corporation. **Soya in food.** Available at: < http://www.cnpso.embrapa.br/soja_alimentacao/index.php?pagina=23>. Accessed 28 Mar 2013.

ESCALANTE, E.E.; WILCOX, J.R. Variation in seed protein among nodes of normal- and high-protein soybean genotypes. **Crop Science,** v. 33, p.1164-1166, 1993.

ESPÍNDOLA, S.M.C.G. et al. **Evaluation of the Oil Content and the Productivity in Soya Strains for Cultivation in the Cerrado.** In: IX SIMPÓSIO NACIONAL CERRADO E II SIMPÓSIO INTERNACIONAL SAVANAS TROPICAIS **(Anais...),** 2008, Brasília. **Challenges and Strategies for the Balance between Society, Agribusiness and Natural Resources.** 2008.

FERRARI, R.A.; OLIVEIRA, V.S.; SACABIO, A. Soybean biodiesel - Conversion rate to ethyl esters, physicochemical characterisation and consumption in an energy generator. **Química Nova,** v.28, n.1, p.19-23, 2005.

GAVIOLI, I.L.C.; LEMOS, L.B.; FARINELLI, R.; CAVARIANI, C. Agronomic and technological performance of conventional soya cultivars with special characteristics for human consumption. **Journal of Agronomic Sciences,** v. 1, p. 84-99, 2012.

GOBIRA, P.S.S.C. **Evaluation of the genetic divergence of soya grown under high and low phosphorus in Tocantins, with a view to biofuel production.** Palmas, TO. Federal University of Tocantins, 2012. 57p. (Master's Degree in Agroenergy) - Federal University of Tocantins, 2012.

GOMES, P. **A soja.** 5th edition. 2nd reprint. São Paulo: Nobel, 1988. p.152.

GONTIA, A.S.; AWASTHI, M.K. Effect of seed grading by size on various seed vigour attributes, morphophysiological characters and seed yield in soybean *[Glycine max* (L.) Merrill] genotypes. **Seed Research,** v.27, n.1, p.333-350, 1999.

HERBINET, O.; PITZ, W.J., WESTBROOK, C.K.; Detailed Chemical kinetic oxidation mechanism for a biodiesel surrogate. **Combustion and Flame,** v.154, n.3 p. 507-528, 2008.

HYMOWITZ, T. On the domestication of the soybean. Economic Botany, 24:408-21, 1970.

HYMOWITZ, T. **Soybeans: The success story.** In: J. Janick and J.E. Simon (eds.), Advances in new crops. Timber Press, Portland, OR., p. 159-163, 1990. Available in: <http://www.hort.purdue.edu/newcrop/proceedings1990/v1-159.html>. Accessed on 20 March 2013.

HUSKEY, L.L.; SNYDER, H.E.; GBUR, E.E. Analysis of single soybean seeds for oil and protein. **Journal of the American Oil Chemists' Society,** v. 67, n.10, p. 165-167, 1990.

IBGE - Brazilian Institute of Geography and Statistics. IBGE Data Recovery System. Available at: < http://www.sidra.ibge.gov.br>. Accessed on 10 Jan 2013.

IRISH, G. G. et al. Removal of the alpha-galctosites of sucrose from soybean meai using either ethanol extraction or exogenous alpha-galactosidase and broiler performance. **Poultry Science,** Saskatoon, v.74, n.9, p. 1484-1494, 1998.

JIANG, H.; EGLI, D.B. Shade induced change in flower and pod number and flower and fruit abscission in soybean. **Agronomy Journal,** Madison, v.85, n.2, p.221-225, 1993.

KNOTHE, G.H.; STEIDLEY, K.R. Cinematic viscosity of biodiesel fuel components and related compounds. Influence of compound structure and comparison to petrodiesel fuel components. **Fuel,** v.84, n.9, p. 1059-1065, 2005.

KRZYZANOWSKI, F.C.; FRANÇA-NETO, J. DE B.; COSTA, N.P. da; HENNING, A.A; VIEIRA, B.G. T.L. Influence of seed size on soya crop productivity. In: SOYBEAN RESEARCH MEETING OF THE CENTRAL REGION OF BRAZIL, 26, 2004, Ribeirão Preto. **Anais-** Londrina: Embrapa Soja, 2004, p.318-319. (Embrapa Soja. Documentos, 234).

KRZYZANOWSKI, F.C.; FRANÇA-NETO, J. DE B.; COSTA, N.P. da; HENNING, A.A.; VIEIRA, B.G. T. L Influence of seed size on soya crop productivity. In: REUNIÃO DE PESQUISA DE SOJA DA REGIÃO CENTRAL DO BRÁSIL, 27, 2005, Cornélio Procópio. **Proceedings...** Londrina: Embrapa Soja, 2004, p.567-568 (Embrapa Soja. Documentos, 257).

KRZYZANOWSKI, F.C.; FRANÇA-NETO, J. DE B.; FARIAS, J.R.B.; COSTA, N.P. da; HENNING, A.A; Influence of seed size on soya bean yield. In: REUNIÃO DE PESQUISA DE SOJA DA REGIÃO CENTRAL DO BRASIL, 30., 2008, Cornélio Procópio. **Proceedings...** Londrina: Embrapa Soja, 2004, p.337-338. (Embrapa Soja. Documentos, 304).

LIMA, A.M.M.P.; CARMONA, R. Influence of seed size on the productive performance of soya beans. **Revista Brasileira de Sementes,** v. 21, n.1, p. 157-163, 1999.

LUI, M.C.Y.; AGUIAR, C.L.; ALENCAR, S.M.; SCAMPARINI, A.R.P.; PARJ, Y.K. Isoflavones in soya protein isolates and concentrates. **Ciência e Tecnologia de Alimentos,** v.23, p.206-212, 2003.

MACHADO, R.P. **Production of ethanol from soya molasses. 151 p. Master's dissertation in Chemical Engineering.** 1999. 151 f. Master's dissertation. Federal University of Rio Grande do Sul, Porto Alegre, 1999.

MANDARINO, J.M.G.; CARRÃO-PANIZZI, M. C. **Soya: the best plan for your health.** Londrina: EMBRAPA-CNPSo. 1997. 6p.

MME. Ministry of Mines and Energy. **Biodiesel.** Online. Available at:< http://www.mme.gov.br/programas/biodiesel/menu/biodiesel/o_biodiesel.htm l>. Accessed on 18 January 2014.

MONYEM, A.; VAN GERPEN, J.H. The effect of biodiesel oxidation on engine performance and emissions. **Biomass and Bioenergy.** v.20, p.317- 325, 2001.

MORAES, L.M.F.; LANA, R.M.Q.; MENDES, C.; MENDES, E.; MONTEIRO, A.; ALVES, J.F. Redistribution of molybdenum applied foliarly at different times in soya bean crop. **Ciência e agrotecnologia,** v.32, n.5, 2008.

MELO, W.M.C.; PINHO, R.G.V.; FERREIRA, D.F. Combinatorial capacity and genetic divergence in commercial maize hybrids. **Ciência e Agrotécnica,** v.25, p.821-830, 2001.

MIRANDA, G.V. **Genetic diversity and performance of soya cultivars as parents.** 1998. 117 p. Thesis (Doctorate in Genetics and Plant Improvement) - Federal University of Viçosa. Viçosa, 1998.

NUNES, G.H.S.; COSTA FILHO, J.H.; SILVA, D.J.H.; CARNEIRO, C.S.; DANTAS, M.S.M. Genetic divergence between lines of melão pele de Sapo. **Revista Ciência Agronómica,** v.42, n.03, p.765-773, 2011.

PANDEY, J.P.; TORRI, E.J.H. Path coefficient analysis of seed yield components in soybean *Glycine max* (L) Merrill. **Crop Science,** v.13, n.5, p.505-507, 1973.

PELÚZIO, J.M., ROCHA, C.R.N., SANTANA, W.; BARROS, H.B. Influence of pod removals on the yield components of soya beans *(Glycine max* (L) Merrill), in Gurupi-TO. **Bioscience Journal,** v.17, n. 1, p.85-96, 2001.

PELÚZIO, J.M.; BARROS, H.B.; ROCHA, R.N.C.; SILVA, R.R.; NASCIMENTO, I.R.; Influence of Artificial Defoliation on Grain Yield and Production Components of Soya. **Ciência Agrotecnologia,** v.26, n.6, p.1197-1203, 2002.

PELÚZIO, J.M.; BARROS, H.B.; BRITO, E.L.; SANTOS, M.M.; SILVA, R.R.; Effects of defoliation at different phenological stages on soya beans. **Revista Ceres,** v. 51, n. 297, p. 575-585, 2004.

PELÚZIO, J.M.; VAZ DE MELO, A.; AFFÉRRI, F.S.; SILVA, R.R.; BARROS, H.B.; NASCIMENTOS, I.R.; FIDELIS, R.R. Genetic variability among soya cultivars under different soil and climate conditions. **Revista Ciência Agronómica,** v.42, n.01, p. 108-115, 2009.

PETERSON, C.L; TABERSKI, J.S.; THOMPSON, J.C.; CHASE, C.L. The effect of biodiesel feedstock on regulated emissions in chassis dynamometer tests of a pickup, ASAE, 1999.

PRATISSOLI, D.; SCHMILDT, E.R; REIS, E.F.; THULER, R.T. Influence of simulated defoliation on productivity and other agronomic characteristics of bean. **Revista Ceres,** v. 48, n. 275, p. 17- 24, 2001.

PROULX, R.A.; NAEVE, S.L. Pod removal, shade, and defoliation effects on soybean yield, protein, and oil. **Agronomy Journal,** v.101, p. 971-978, 2009.

RAO, C.R. **Advanced statistical methods in biometric research.** New York: John Willey, 1952, 390 p.

FUNCTIONAL AND NUTRACEUTICAL MAGAZINES. ***Soya: history, trends and virtues.*** Editora Insumos, p. 28-40. São Paulo - SP, 2007. Available at< http://www.insumos.com.br/funcionais_e_nutraceuticos/materias/76.pdf>. Accessed 31 Mar 2013.

ROCHA, V.S.; SEDIYAMA, T.; SEDIYAMA, C.S.; SILVA, R.F.; SEDIYAMA, T.; Influence of pod removal on soya bean yield components. **Revista Ceres,** v.43, n. 246, p. 126-138,1996.

ROCHA, M.C; GONÇALVES, L.S.A.; RODRIGUES, R.; SILVA, P.R.A.; CARMO, M.G.F.; ABBOUD, A.C.S. Use of the Gower algorithm to determine genetic divergence between cherry tomato accessions. **Acta Scientiarum. Agronomy,** v. 32, n. 3, p. 423-431,2010.

SALES, P.V.G.; PELUZIO, J.M.; AFFÉRRI, F.S.; SILVA, M.C.C.; SALES, V.H.G. Variability of pod position in terms of oil content in soya beans. **Revista Ciências Agrárias (Belém),** v. 56, n. 3, p. 274-277, 2013.

SANDUN, F.; KARRA, P.; HERNANDEZ, R.; JHA, S.K. **Effect of incommpletely converted soybean oil on biodiesel quality.** Elsevier Ltd. All rights reserved, 2005.

SANTOS, M.A.; **Insertion of biodiesel in the Brazilian energy matrix: Technical and environmental aspects related to its use in combustion engines.** Dissertation Inter-University Graduate Programme in Energy, University of São Paulo, University of São Paulo, 118 p. 2007.

SANTOS, E.R.S.; BARROS, H.B.; FERRAZ, E.C.; CELLA, A.J.S.; CAPONE, A.; SANTOS, A.F.; FIDELIS, R.R. Divergence between soya genotypes grown in irrigated floodplains. **Revista Ceres,** v.58, n.6, 2011.

SANTOS, E. R.; SANTOS, A. F.; CAPONE, A.; SANTOS, W. R.; MOURA, S. G.; BARROS, H. B. Genetic dissimilarity between soybean genotypes cultivated in lowland irrigated inter-cropping. **Journal of Biotechnology and Biodiversity,** v.4, n.3, p.222-231,2013.

SCAPIM, C.A.; PIRES, I.E.; CRUZ, C.D.; AMARAL JÚNIOR, A.T.; BRACCINI, A.L.; OLIVEIRA, V.R. Evaluation of genetic diversity in Eucalyptus camaldulensis Dehnh, by means of multivariate analysis. **Revista Ceres,** v.6, p.347-356, 1999.

SEDIYAMA, T.; TEIXEIRA, R.C.; REIS, M.S. **Improving soya beans.** In: BORÉM, A. Improvement of cultivated species. Viçosa: UFV, 2005. 969 p.

SEDIYAMA, T.; **Production technology and uses of soya.** Londrina: Ed. Mecenas, p.313, 2009.

SILVA, J.M. da; LIMA, J.S. de S.; AGUIAR, A.V.; Ramos, K.S.; ZUCOLOTO, M. Determination of the Need for Soya Bean Liming in Two Soil Preparation Systems in Cerrado Areas. **Anais...** X Encontro Latino Americano de Iniciação Científica e VI Encontro Latino Americano de Pés-Graduação, 2006, São Jose dos Campos. UNIVAP MAGAZINE.

SOUZA, L.C.F.; ZANON, G.D.; PEDROSO, F; ANDRADE, L.H.L. Protein and oil content in soya beans as a function of seed treatment and micronutrient application. **Ciência e Agrotecnologia.** v. 33, n. 6, p. 1586-1593, 2009.

SUDRÉ, C.P.; GONÇALVES, L.S.A.; RODRIGUES, R.; AMARAL JÚNIOR,

A.T.; RIVA-SOUZA, E.M.; BENTO, C.S. Genetic variability in domesticated Capsicum spp. as assessed by morphological and agronomic data in mixed statistical analysis. **Genetics and Molecular Research,** v. 9, n. 1, p. 283- 294, 2007.

TOLEDO, M.Z.; GARCIA, R.A.; PEREIRA, M.R.R.; BOARO, C.S.F.; LIMA, G.P.P. Nodulation and Nitrate Reductase Activity as a Function of Molybdenum Application in Soybean. **Bioscience journal,** v. 26, n. 6, p.858-864, 2010.

VELLO, N. A.; SILVA, L. A. S. Genetics seeks to meet growing human consumption. **Visão Agrícola,** year 3, n.5, p.60 - 62, 2006.

VEIGA, A. D.; VON PINHO, E.V.R.; VEIGA, A.D.; PEREIRA, P.H.A.R.; OLIVEIRA, K.C.; VON PINHO, R.G. Influence of potassium and liming on the chemical composition, physiological quality and enzymatic activity of soya beans. **Ciência e Agrotecnologia,** v.34, n.4, 2010.

VENDRAMINI, J.M.; NEVES, L.G.; BARELLI, M.A.A.; ELIAS, J.C.F.; LUZ, P.B. Optimising the use of cassava morpho-agronomic descriptors in multivariate analysis. Revista Ciência Agronómica, v.42, n.04, p.906-913, 2011.

VIANA, C.F.A.; SILVA, M.A. e; PIRES, A.V.; LOPES, P.S.; PIASSI, M. Estudo da divergência genética entre quatro linhagens **de** matrizes **de frango de** corte utilizando técnicas **de** análise multivariada, **Revista Brasileira de Zootecnia,** v.29, n.4, p.1074-1081,2000.

VILELA, F.O.; AMARAL JÚNIOR, A.T.; PEREIRA, M.G.; SCAPIM, C.A.; VIANA, A.P.; FREITAS JÚNIOR, S.P. Effect of recurrent selection on the genetic variability of the UNB-2U popcorn population using RAPD markers. **Acta Scientiarum. Agronomy,** v. 30, n. 1, p. 25-30, 2008.

VOGT, G.A.; BALBINOT JÚNIOR, A.A.; SOUZA, A.M. Genetic divergence between sunflower cultivars in the northern plateau of Santa Catarina. **Scientia Agraria,** v.11, n.04, p.307-315, 2010.

ZANON, G.D. **Protein and Oil Content in Soya Beans Under Different Types of Management.** 2007. 62 f. Thesis (Doctorate) - Federal University of Grande Dourados, Dourados, Ms, 2007.

CHAPTER 2

OIL AND PROTEIN CONTENT IN SOYA BEANS GROWN AT LOW LATITUDE IN DIFFERENT PLANT POSITIONS.

INTRODUCTION

Commercial interest in the crop is related to its high protein content, averaging 40 per cent, oil content, averaging 20 per cent, and its productivity (BONATO et al, 2000; SEDIYAMA et al, 2005; VELLO & SILVA, 2006). The grain also contains an average of 34% carbohydrates and 5% minerals (EMBRAPA, 2013). As well as being an important food source of protein and oil, it is also rich in minerals such as iron, calcium, phosphorus, potassium and B vitamins (GAVIOLI et al., 2012).

In addition, soya has bioactive compounds such as isoflavones, which have been widely studied for their beneficial effects on human health in the control of cancer, cardiovascular problems, osteoporosis, cholesterol, diabetes, obesity and climacteric syndrome (MANDARINO & CARRÃO-PANIZZI, 1997; LUI et aL, 2003; GAVIOLI, 2012).

The chemical composition of soya is considerably influenced by the genetics of the cultivars (MACHADO, 1999; ÁVILA et aL, 2007; SOUZA et aL, 2009) and the climatic conditions in which the plants grow (MACHADO, 1999, ALBRECHT et aL, 2008; BARBOSA et aL, 2011).

Other factors that influence the protein and oil composition of soya beans are fertiliser levels (MORAES et al., 2008; TOLEDO et al., 2010, VEIGA et al., 2010), shading (PROULX & NAEVE, 2009; BELLALOUI & GILLEN, 2010), pod removal (ROCHA et al. 1996; PELUZIO et al. 2001; PROULX & NAEVE, 2009), leaf removal (CAMPELO et al. 1999; PRATISSOLI et al. 2001; PELUZIO et al. 2002; PELUZIO et al. 2004; PROULX & NAEVE, 2009), position of the pods at the nodes of the main stem (COLLINS & CARTTER, 1956; HUSKEY et al.,1990ESCALANTE & WILCOX, 1993; BENETT et al., 2003; BELLALOUI & GILLEN, 2010; SALES et al., 2013).

In relation to the position of the pods on the nodes of the main stem of the plant, Benett et al. (2003), in soya cultivation under field conditions, Bellaloui & Gillen (2010), in cultivation under greenhouse conditions, both carried out in the United States, and Sales et al. (2013), in trials carried out under field conditions in Brazil, in the city of Palmas, state of Tocantins, observed a trend towards a higher concentration of oil content in the beans located in the lower third of the plants and a higher concentration of proteins in the upper third. On the other hand, in research carried out by Huskey et al. (1990), no significant difference was observed in the protein content of soya beans between the upper and lower thirds of the plant, which in turn were higher than the middle third.

In view of the above, variations in the oil and protein content of the grains in the different parts of the plant can result in errors in quantifying these characteristics during the sampling process, which is extremely important in breeding programmes. This study was carried out with a view to selecting genotypes with a higher oil content for biofuel production and a higher protein content for human food and feed production.

MATERIALS AND METHODS

Two experiments were carried out in the experimental area of the Federal University of Tocantins (UFT) in the municipality of Palmas - TO, located at latitude $10^{s}45'$ S, longitude $47^{e}14'$ W at an altitude of 220m. The climate of the region, according to the Kõppen classification, is of the Aw type, tropical with a dry season. The first experiment was installed in the 2011/12 agricultural year (December 2011) and the second in the 2012/13 agricultural year (December 2012). The chemical characteristics of the soil (Latossolo Vermelho Amarelo Distrófico) at a depth of 0-20 cm in the two agricultural years are shown in Table 11-1.

Table 11-1. Soil analysis results from the Palmas experimental field in the 2011/12 and 2012/13 crop years.

Harvest	pH	K (mg	P* (mg dm-	M.O.	CTC	S. B	Clay	Silt	Sand	Classification

years		dnrr³)	³)			(%)	(%)	(%)	(%)	
2011/12	4,9	42	25,5	1,8	3,08	79,80	20	8	72	Sandy loam
2012/13	5,8	95	18,3	1,4	5,54	71,12	23	6	71	Sandy loam

* Melich Method; O.M.: Organic Matter, CEC: Cation Exchange Capacity, B.S.: Base Saturation.

The experimental design used was randomised blocks, with 90 treatments and three replications. The treatments were arranged in a sub-subdivided plot design, with the agricultural years (2011/12 and 2012/13) allocated to the plots and 15 soya cultivars allocated to the sub-plots [BRS Valiosa, P98Y51, P98Y70, P99R03, M8527RR, M8925RR, M9144RR, M8867RR, TMG115RR (transgenic), Conquista, Tucunaré, Pintado, M8866, M9350 and TMG803 (conventional)], and in the subplots the positioning of the pods on the main stem of the plant (lower third, middle third and upper third). These cultivars were chosen because they are widely used in the soil and climate conditions of the central region of the state of Tocantins.

The experimental plot consisted of four rows of 5.0m long soya plants, spaced 0.45m apart. When harvesting, only plants from the two centre rows were used, discarding 0.50m from the end of each row.

The soil was prepared by ploughing, harrowing and furrowing in both agricultural years. Sowing fertiliser, carried out after soil analysis and according to the crop's requirements, was 400 kg ha^{1} of the NPK 05-25-15 formulation, followed by top dressing at stage R2 (full flowering) with 100 kg ha^{1} of K2O, using potassium chloride as the source.

The sowing density was aimed at obtaining 14 plants per linear metre, and thinning was carried out when necessary at 10 days after emergence. At the time of sowing, the beans were treated with fungicides (Carboxin + Thiram 200 SC -> 2.5mL + 2.5mL of water kg^{1}), followed by inoculation of the beans with *Bradyrhizobium japonicum* strains, using the product Biomax® Premium Turfa - Soja (Strains 5079 and 5080), at a dosage of 60 g 50 kg^{1} of soya beans.

Pest, disease and weed control was carried out as necessary. Methamidophos (300 g $ha^{1)}$ was used to control pests at the Fte, FU and Re stages. To control Asian rust, two applications of the fungicide Tebuconazole

200 EC (0.5 L ha$^{1)}$ were made, one at R2 and the other at R4. Weeds were controlled with one application of Roundup Ready, at stages Ve and R2 of the soya bean, at a dose of 1.44 L ha'^{1}.

In each experimental plot, 30 plants of each cultivar were harvested from the two centre rows, discarding 0.50 m from the ends, and divided into the upper, middle and lower thirds, as shown in Figure 11-1.

Next, the 30 pods from each position/plot were threshed and the grains packed into a single paper bag, which was identified by cultivar and pod position, and transported to the Agricultural Research Laboratory-LPA of the Food Engineering Course at the Federal University of Tocantins - Palmas Campus, where the grains were milled.

The protein content on a dry basis was obtained using the Kjedhal method, converting the % of N into % of protein using a factor equal to 6.25 for cereals and oilseeds, according to the Adolfo Lutz Institute (IAL, 2005). The dry lipid content for each cultivar in each third of the plant was determined using the Soxhlet method (IAL, 2005).

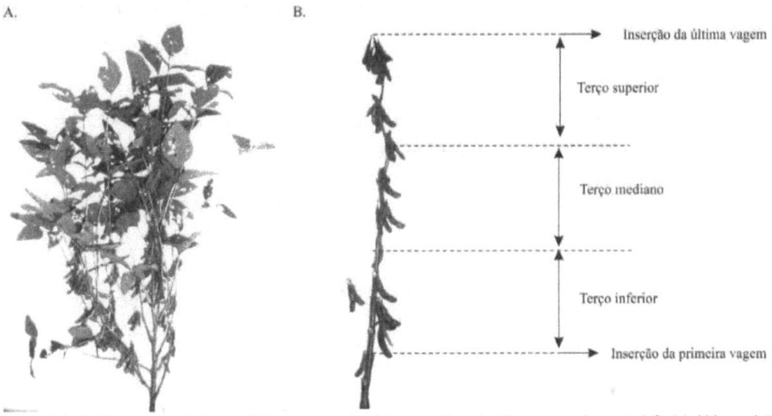

Figure 11-1. Representation of the soya plant harvested in the experimental field (A) and the section of the plant into thirds (lower, middle and upper) from the insertion of the first and last pod of the plant (B).

The protein and oil content data were subjected to the Kolmogorov-Smirnov normality test using the ASSISTAT Version 7.6 beta statistical programme (SILVA & AZEVEDO, 2002).

Next, the characteristics were analysed for variance and the means were compared using the Scott-Knott test at a 5% significance level, which according to Borges & Ferreira (2003) is more objective and clearer in interpreting the data, with no ambiguity.

RESULTS AND DISCUSSIONS

The analysis of variance for protein and oil content showed a significant effect for cultivar, pod position and cultivar x pod position interaction (Table II-2).

The significant effect of cultivars demonstrates the existence of variability between cultivars. The significant cultivar x pod position interaction indicates the different behaviour of the cultivars in terms of pod position. Therefore, the splits were carried out. The coefficients of variation were low, indicating the good precision of the experiment.

The protein content data had a normal distribution ($p > 0.05$), according to Kolmogorov-Smirnov (1933), cited by Lilliefors (1967), , which meant that the data did not need to be transformed to carry out the ANOVA. This was not the case for the oil content, which was transformed using the Shapiro-Wilk method, using the transformed x = x.√

Table II-2. Summary of the analysis of variance for the characteristics oil content (%) and protein (%) in 15 soya cultivars, in different parts of the plant, 2011/12 and 2012/13 harvests.

FV	GL	MIDDLE SQUARE	
		PROT	OIL[1]
Blocks	2	0.334^{ns}	0.016^{ns}
Agricultural Year (a)	1	0.123^{ns}	0.137^{ns}
Waste (a)	2	0,4757	0,0103
Cultivars (b)	14	31,46**	0,872**
Year X Cultivars	14	19,36"	0,047**
Residue (b)	56	0,610	0,00464
Position (c)	2	206,17**	0,258"
Year x Position	2	3,51**	0,021"
Cultivars X Position	28	5,77**	0,103**
Year X Cultivars X Position	28	2,27**	0,011**
Residue (c)	120	0,499	0,0043
	MEDIA	38,37	20,89

CV%(a)	1,80	2,23
CV%(b)	2,04	1,49
CV%(c)	1,84	1,43

[1] Data transformed by the Shapiro-Wilk method, using the transformed x =√ x.
: significant at the 1% significance level (p < 0.01); ns: not significant (p >= 0.05). **(PROT): Grain protein, in %; **(OIL):** Oil content, in %.

Protein content of grains

The averages for the protein content of the cultivars in the different plant positions are shown in (Table II-3).

Table 11-3. Average protein percentage of 15 soya cultivars in Palmas, Tocantins, 2011/12 and 2012/13 harvests.

CULTIVARS	LOWER THIRD	MIDDLE THIRD	UPPER THIRD	AVERAGE
BRS Valiosa	38.25 aB	38.04 dB	40.08 dA	**38,79 c**
TMG803	34.93 cC	38.04 dB	40.76 cA	**37,92 d**
M8866	36.92 bB	37.55 dB	38.61 eA	**37,69 d**
P98Y51	35.67 cC	37.39 dB	39.60 dA	**37,55 d**
M8925RR	37.81 bB	40.54 aA	40.76 cA	**39,70 b**
M8867RR	38.74 aB	38.73 cB	39.69 dA	**39,05 c**
P98Y70	37.29 bC	39.41 bB	41.40 bA	**39,37 b**
TMG115RR	37.21 bB	37.85 dB	38.59 eA	**37,89 d**
P99R03	37.33 bA	37.46 dA	37.78 eA	**37,52 d**
M9144RR	34.85 cC	38.61 cB	41.07 cA	**38,18 d**
Painted	36.55 bC	39.57 bB	42.06 aA	**39,39 b**
M8527RR	37.34 bC	39.52 bB	40.42 cA	**39,09 c**
Achievement	37.25 bB	37.77 dB	38.35 eA	**37,79 d**
Tucunaré	38.78 aC	40.62 aB	42.53 aA	**40,65 a**
M9350	33.38 dB	35.49 eA	35.99 fA	**34,96 e**
AVERAGE	**36,82 C**	**38,44 B**	**39,85 A**	

Averages followed by the same uppercase letter in the row and lowercase letter in the column do not differ at a 5% significance level using the Scott-Knott test.

With the exception of P99R03, which showed no significant difference in protein content in grains located in different parts of the plant, the other cultivars showed a higher protein content in grains located in the upper third, followed by the middle third. These results are in agreement with those obtained by Sales et al. (2012), who also found a higher protein content in grains located in the middle and apical regions of the plants in a trial involving soya beans in the state of Tocantins.

According to Harper (1987), the plant's N acquisition system involves two interdependent routes: 1) the absorption and reduction of NO3 by the nitrate reductase enzyme and; 2) the fixation of N2 by *Bradyrhizobium japonicum* in the nodules with the help of the nitrogenase enzyme, both of which depend on the energy provided by photosynthesis during the grain filling period (HARPER, 1987; PROULX & NAEVE, 2009). Fixation is dependent on photosynthesis which, according to Burton (1994), will have a positive influence on oil content, protein and seed production.

Egli & Bruening (2007) suggest that the supply of total nitrogen to the plant is the main limiting factor for the accumulation and yield of protein in soya beans. Bellaloui & Gillen (2010) suggest that N mobilisation in the seed is more dependent on the amount of N accumulated in plant tissues than on the amount supplied by the soil during grain filling.

According to studies by Kim et al. (1999) and Hirai et al. (2002) there is experimental evidence suggesting the involvement of specific metabolites in protein accumulation in seeds. The concentration of O-acetylserine (OAS), which is an intermediate metabolite in the synthesis of cysteine, plays an important role in the accumulation of 7S and 11S proteins, and this accumulation is due to sulphur deficiency (KIM et aL, 1999). It appears that OAS coordinates the signal from the availability of photoassimilates and the N/S ratio. As the N/S ratio is higher in the upper nodes of the plant, there is an accumulation of β-conglycine subunits, promoting a higher protein yield in these nodes (BENNETT et al., 2003).

In this sense, the lower concentration of proteins in the grains located in the lower third can be explained by the reduction in the activities of the enzymes nitrate reductase, nitrogenase and the N/S ratio, due to the lower photosynthetic activity of the lower leaves, as a result of shading (BELLALOUI & GILLEN, 2010), as well as the senescence of the basal leaves (NAOE et aL, 2006). On the other hand, in the middle and apical thirds of the plant there is greater photosynthetic activity, resulting in intense translocation of photoassimilates and nitrogen fixation (NELSON, 1961), as well as greater accumulation of OAS (KIM et al., 1999; HIRAI et al., 2002; BENNETT et al., 2003) which will lead to an increase in the N/S ratio and, consequently, an increase in protein.

An average reduction of 3.03% in protein content was detected between the upper third (39.85%) and the lower third (36.82%). These results were similar to those found by Collins & Cárter (1956), Bennett et al., (2003), Naoe et al., (2006), Bellalou & Gillen (2010) and Sales (2012), who observed a reduction between these thirds of 1.0%, 4.0%, 5.7%, 2.48% and 3.03% respectively. On the other hand, Huskey et al. (1990) observed no significant differences in protein concentration between the upper and lower thirds.

Significant differences were detected between the cultivars, within each position, for protein content, represented by different groups (Table I-3). Six groups of averages were formed for the upper third. The group with the highest average was represented by Tucunaré (42.5%) and Pintado (42.06%), followed by the group represented only by cultivar P98Y70 (41.4%). The M9350 cultivar (35.99%) alone had the lowest protein content.

In the middle and lower thirds, five and four groups of averages were formed, respectively. For these regions, the Tucunaré cultivar achieved the highest protein value with 44.62% and 38.78% respectively, without, however, differing significantly from M8925RR (40.54%) for the middle third, and M8867RR (38.74%) and BRS Valiosa (38.25%) for the basal region. For both thirds, the lowest protein content in the grains came from M9350.

It should be noted that the Tucunaré and M9350 cultivars had the highest and lowest protein content in the grains, respectively, regardless of the position of the grains.

There was a variation of 34.96 to 40.65% in the protein content of the cultivars studied. These values are in line with those observed by Bennett et al. (2003), Zanon (2007), Bellaloui & Gillen (2010) and Sales (2012), who worked with different cultivars that showed variations of 30 to 48.6%.

Grain oil content

The averages for the oil content of the cultivars in the different plant positions are shown in (Table 11-4). Most of the cultivars, with the exception of BRS Valiosa, showed significant differences in oil content for the different positions on the plant, with the highest average values being observed in the lower and middle thirds. In their studies, Bennett et al. (2003), Bellaloui & Gillen (2010) and Sales et al. (2013), the latter carried out in the state of Tocantins, also found a lower concentration of oil in soya beans in the pods in the upper third.

Table II-4. Average oil content of 15 soya cultivars in Palmas, Tocantins, 2011/12 and 2012/13 harvests.

CULTIVARS	LOWER THIRD	MIDDLE THIRD	UPPER THIRD	AVERAGE
BRS Valiosa	22.21 cA	21.74 cA	21.80 bA	**21,92 c**
TMG803	23.74 bA	21.61 cB	20.79 cC	**22,05 C**
M8866	20.10 eA	20.68 dA	18.99 dB	**19,92 e**
P98Y51	24.16 bA	22.14 bB	21.94 bB	**22,74 b**
M8925RR	22.77 cA	20.74 dB	21.18cB	**21,57 d**
M8867RR	16.92 gB	15.27 gC	17.89eA	**16,69 g**
P98Y70	21.00 dA	19.19eB	19.47 dB	**19,88e**
TMG115RR	22.63 cA	22.66 bA	20.99 cB	**22,09 c**
P99R03	22.35 cB	23.65 aA	24.01 aA	**23,34 a**
M9144RR	18.15 fC	21.49 cA	20.41 cB	**20,02 e**
Painted	21.09 dB	22.62 bA	21.97 bA	**21,89 c**
M8527RR	17.39 gA	17.46fA	16.40 fB	**17,08 f**
Achievement	22.66 cA	20.13 dC	21.43 bB	**21,40 d**
Tucunaré	21.36 dA	19.29 eB	19.38 dB	**20,01 e**
M9350	24.80 aA	23.48 aB	19.77 dC	**22,68 b**
AVERAGE	**21,42 A**	**20,81 B**	**20,43 C**	

Averages followed by the same uppercase letter in the row and lowercase letter in the column do not differ at a 5% significance level using the Scott-Knott test.

It should be noted, however, that the cultivars M8867RR, P99R03, Pintado and M9144RR had higher average oil content in the grains of the upper and/or middle third.

The difference in the response of the cultivars to the average oil content in the grains along the plant stem may be associated with plant architecture, growth habit (indeterminate and determinate), maturity (Bellaloui & Gillen,

2010) and shading (Burkley et aL, 1997).

According to Burkley et al. (1997), under high population density, soya beans in the basal position of the plant had 25% more lipids compared to the upper part, which was accompanied by a proportional increase in oleic and linoleic acids. Finally, they concluded that the accumulation of triacylglycerol was a result of altered carbon metabolism and not a response to senescence.

On average, a percentage reduction of 0.99% was detected in the oil content of grains from the top third of the plant (20.43%), when compared to the bottom third of the plant (21.42%). These results are well below those reported by Bellaloui & Gillen (2010), Sales (2013) who found a reduction of 4.9% and 4.4%, respectively.

Significant differences were detected between the cultivars, within each position, for oil content (Table I-4).

The average oil content of the cultivars evaluated ranged from 15.27% (M8867RR) to 24.80% (M9350). These results are close to those obtained by Bellaloui & Gillen (2010), Barbosa et al. (2011) and Sales (2013), who found variations of 17.31 to 24.69% between the cultivars, the latter two having been carried out in the state of Tocantins.

The lower and middle thirds of the main stem had seven groups of averages. In the lower third, the group with the highest average was made up only of M9350 (24.80%), followed by P98Y51 (24.16%) and TMG103RR (23.74%). The cultivars M8527RR (17.39%) and M8867RR (16.92%) had the lowest average oil content in the grains. For the middle third, cultivar M9350 once again had the highest oil content (23.48%), although it did not differ significantly from P99R03 (23.65%). Similarly to the middle third, M8867RR had the lowest oil content (15.27%).

For the upper third of the main stem of the plant, six groups of averages were formed for oil content. Cultivar P99R03 had the highest average (24.01%), differing significantly from the averages of Pintado (21.97%), P98Y51 (21.94%), Conquista (21.43%) and BRS Valiosa (21.80%), the latter essentially derived from Conquista for the RR gene. The group with the lowest average was made up only of M8527RR (16.40%).

CONCLUSIONS

There were differences between the soya cultivars in terms of protein and oil content.

The grains located in the upper third of the main stem have a higher percentage of protein and the grains located in the middle and lower thirds of the main stem have a higher percentage of oil;

In plant breeding programmes aimed at selecting superior genotypes for energy production, grain sampling should be carried out on grains from pods located in the same position on the plant.

BIBLIOGRAPHICAL REFERENCES

ALBRECHT, L.P.; BRACCINI, A.L.; ÁVILA, M.R.; SUZUKI, L.S.; SCAPIM, C.A.; BARBOSA, M.C. Oil and protein content and yield of soya beans as a function of early sowing in western Paraná. **Bragantia,** v. 67, n. 04, p. 865-873, 2008.

ÁVILA, M.R.; BRACCINI, A.L.; SCAPIM, C.A.; MANDARINO, J.M.G.; ALBRECHT, L.P.; FILHO, P.S.V. Yield Components, Isoflavone, Protein and Oil Content and Quality of Soya Seeds. **Revista Brasileira de Sementes,** v. 29, n. 3, p.111-127, 2007.

BARBOSA, V.S.; PELUZIO, J.M.; AFFÉRRI, F.S.; SIQUEIRAS, G.B. Behaviour of soya cultivars at different sowing times for biofuel production. **Revista Ciência Agronómica,** v.42, n.3, p.742-749, 2011.

BELLALOUI, N; GILLEN, A.M.; Soybean seed protein, oil, fatty acids, N, and S partitioning as affected by node position and cultivar differences. **Agricultural Sciences,** v.1, n.3, p.110-118, 2010.

BENNETT, J.O.; KRISHNAN, A.H.; WIEBOLD, W.J.; KRISHNAN, H.B. Positional Effect on Protein and Oil Content and Composition of Soybeans. **Agricultural and Food Chemistry,** v.51, p.6882-6883, 2003.

BONATO, E.R.; BERTAGNOLLI, P.F.; LANGE, C.E.; RUBIN, S. de A.L.Oil and protein content in soya genotypes developed after 1990. **Pesquisa Agropecuária Brasileira,** v.35, p.2391-2398, 2000.

BORGES, L.C.; FERREIRA, D.F. Power and type i error rates of the Scott-Knott, Tukey and student-newmankeuls tests under normal and non-normal residue distributions. **Revista Matemática e estatística,** v.21, n.1, p.67-83, 2003.

BURKLEY, K.O.; WILSON, R.F; WELLS, R. Effects of canopy shade on the lipid composition of soybean leaves. **Physiologia Planarum,** v.101, p.591-598, 1997.

BURTON, J.W. Breeding soybeans for improved protein quantity and quality. In: WORLD SOYBEAN RESEARCH CONFERENCE, 3., 1984, Ames. **Proceedings...** Boulder: Westview, 1985. p. 361-367.

CAMPELO, J.E.G.; SEDIYAMA, T.; ROCHA, V.S.; SEDIYAMA, C.S.; Effects of defoliation on the onset of flowering in winter soya beans. **Revista Ceres,** v.46, n.268, p.652-666, 1999.

COLLINS, F.I.; CARTTER, J.L.; Variability in Chemical composition of seed from different portions of the soybean plant. **Agronomy Journal,** v.48, p.216-219, 1956.

EGLI, D.B.; BRUENING, W.P. Nitrogen accumulation and redistribution in soybean genotypes with variation in seed protein concentration. **Plant and Soil,** v.301, p.165-172, 2007.

EMBRAPA - Brazilian Agricultural Research Corporation. National Soil Research Centre. **Brazilian soil classification system.** Rio de Janeiro, 2006. 306p.

EMBRAPA - Brazilian Agricultural Research Corporation. **Soya in food.** Available at: <http://www.cnpso.embrapa.br/soja_alimentacao/index.php?pagina=23>. Accessed on 28 March 2013.

ESCALANTE, E.E.; WILCOX, J.R. Variation in seed protein among nodes of normal- and high-protein soybean genotypes. **Crop Science,** v.33, p.1164-1166, 1993.

GAVIOLI, I.L.C.; LEMOS, L.B.; FARINELLI, R.; CAVARIANI, C. Agronomic and technological performance of conventional soya cultivars with special characteristics for human consumption. **Journal of Agronomic Sciences,** v.1, p.84-99, 2012.

HARPER, J.E. Nitrogen metabolism. *In* J.R. Wilcox (ed.) Soybeans: Improvement, production, and uses. 2nd ed. ASA, CSSA, and SSSA, Madison, WL 1987. p.497-533.

HIRAI, M.Y.; KIM, H.; HAYASHI, H.; CHINO, M.; SATOSHI, N.; FUJIWARA, T. Independent roles of methionine and O-acetyl-L-serine in the regulation of the p-subunit gene of p-conglycinin. **Soil Science Plant Nutrition,** v.48, p.87-94, 2002.

HUSKEY, L.L., Snyder, H.E. and Gbur, E.E. Analysis of single soybean seeds for oil and protein. **Journal of the American Oil Chemists' Society,** v.67, n.10, p.165-167, 1990.

ADOLFO LUTZ INSTITUTE. **Analytical standards:** chemical and physical methods for food analysis. 4. ed. São Paulo, 2005.1018 p.

KIM, H.; HIRAI, M.Y.; HAYASHI, H.; CHINO, M.; NAITO, S.; FUJIWARA, T. Role of O-acetyl-L-serine in the coordinated regulation of the expression of a soybean seed storage-proteingene by sulfur and nitrogen nutrition. **Planta,** v.209, p.282-289, 1999.

LILLIEFORS, H.W. On the Kolmogorov-Smirnov Test for Normality with Mean and Variance Unknown. **Journal of the American Statistical Association,** v.62, p.399-402, 1967.

LUI, M.C.Y.; AGUIAR, C.L.; ALENCAR, S.M.; SCAMPARINI, A.R.P.; PARJ, Y.K. Isoflavones in soya protein isolates and concentrates. **Ciência e Tecnologia de Alimentos,** v.23, p.206-212, 2003.

MACHADO, R.P. **Production of ethanol from soya molasses. 151 p. Master's dissertation in Chemical Engineering.** 1999. 151 f. Master's dissertation. Federal University of Rio Grande do Sul, Porto Alegre, 1999.

MANDARINO, J.M.G.; CARRÃO-PANIZZI, M.C. **Soya: the best plan for your health.** Londrina: EMBRAPA-CNPSo. 1997. 6p.

MORAES, L.M.F.; LANA, R.M.Q.; MENDES, C.; MENDES, E.; MONTEIRO, A.; ALVES, J.F. Redistribution of molybdenum applied foliarly at different times in soya bean. **Ciência e agrotecnologia,** v.32, n.5, 2008.

NAOE, L.K.; SEDIYAMA, C.S.; PIOVESAN, N.D.; COIMBRA, R.R.; MOREIRA, M.A. Evaluation of protein content in seeds from different parts of the soya plant. **Revista Ciência Agroambiental,** v.1, n.10, p.69- 73, 2006.

NELSON, C.D.; CLAUSS, H.; NORTIMER, D.C.; GORHAN, P.R. Select translocation of products of photosyntesis on the soybean. **Plant Physiology,** v.36, n.35, p.581-588, 1961.

PELÚZIO, J.M., ROCHA, C.R.N., SANTANA, W.; BARROS, H.B. Influence of pod removal on the yield components of soya beans *(Glycine max* (L) Merrill), in Gurupi-TO. **Bioscience Journal,** v.17, n.1, p.85-96, 2001.

PELÚZIO, J.M.; BARROS, H.B.; ROCHA, R.N. C.; SILVA, R.R.; NASCIMENTO, LR.; Influence of Artificial Defoliation on Grain Yield and Production Components of Soya. **Ciência e Agrotecnologia,** v.26, n.6, p.1197-1203, 2002.

PELÚZIO, J.M.; BARROS, H.B.; BRITO, E.L.; SANTOS, M.M.; SILVA, R.R.;

Effects of defoliation at different phenological stages on soya beans. **Revista Ceres,** v.51, n.297, p.575-585, 2004.

PRATISSOLI, D.; SCHMILDT, E. R; REIS, E. F.; THULER, R. T. Influence of simulated defoliation on productivity and other agronomic characteristics of bean. **Revista Ceres,** v.48, n.275, p.17- 24, 2001.

PROULX, R.A.; NAEVE, S.L. Pod removal, shade, and defoliation effects on soybean yield, protein, and oiL **Agronomy Journal,** v.101, p.971-978, 2009.

ROCHA, V.S.; SEDIYAMA, T.; SEDIYAMA, C.S.; SILVA, R.F.; SEDIYAMA, T.; Influence of pod removal on soya bean yield components. **Revista Ceres,** v.43, n.246, p.126-138, 1996.

SALES, P.V.G. **Genetic variability and the effect of pod position on oil and protein content in soya bean plants for breeding purposes.** 2012. 33 f. Master's thesis. Federal University of Tocantins, Palmas, 2012.

SALES, P.V.G.; PELÚZIO, J.M.; AFFÉRRI, F.S.; SILVA, M.C.C.; SALES, V.H.G. Variability of pod position in terms of oil content in soya beans. **Revista Ciências Agrárias (Belém),** v.56, n.3, p.274-277, 2013.

SEDIYAMA, T.; TEIXEIRA, R.C.; REIS, M.S (2005). **Improving soya beans.** In: BORÉM, A. Melhoramento de espécies cultivadas. Viçosa: UFV, 2005. 969 p.

SILVA, F.A.S.; AZEVEDO, C.A.V. Version of the Assistat computer programme for the Windows operating system, 2002.

SOUZA, L.C. F.; ZANON, G.D.; PEDROSO, F; ANDRADE, L.H.L. Protein and oil content in soya beans as a function of seed treatment and micronutrient application. **Ciência e Agrotecnologia,** v.33, n.6, p. 1586-1593, 2009.

TOLEDO, M.Z.; GARCIA, R.A.; PEREIRA, M.R.R.; BOARO, C.S.F.; LIMA, G.P.P. Nodulation and Nitrate Reductase Activity as a Function of Molybdenum Application in Soybean. **Bioscience journal,** v. 26, n. 6, p.858-864, 2010.

VELLO, N.A.; SILVA, L.A.S. Genetics seeks to meet growing human consumption. **Visão Agrícola,** year 3, n.5, p.60 - 62, 2006.

VEIGA, A. D.; VON PINHO, E.V.R.; VEIGA, A.D.; PEREIRA, P.H.A.R.; OLIVEIRA, K.C.; VON PINHO, R.G. Influence of potassium and liming on the chemical composition, physiological quality and enzymatic activity of soya beans. **Ciência e Agrotecnologia,** v.34, n.4, p.953-960, 2010.

VOLLMANN, J.; FRITZ, C.N., WAGENTRISTL, H.; RUCKENBAUER, P. Environmental and genetic variation of soybean seed protein content under Central European growing conditions. **Journal Science Food Agriculture,** London, v.80, p.1300-1306, 2000.

ZANON, G.D. **Protein and Oil Content in Soya Beans Under Different Types of Management.** 2007. 62 f. Thesis (Doctorate) - Federal University of Grande Dourados, Dourados, MS, 2007.

CHAPTER 3

GENETIC DIVERGENCE IN DIFFERENT PLANT PARTS OF SOYA CULTIVARS ADAPTED TO LOW LATITUDE CONDITIONS.

INTRODUCTION

According to Edmar et al. (2008) and Clemente & Cahoon (2009), soya is the most important oilseed crop grown in the world. However, it was only at the beginning of the last century that it took centre stage in the Western world (ZANON et al, 2007; GOMES et al, 1988; BONETT et al, 1999).

According to Bonato et al. (2000), Sedyama et al. (2005) and Vello & Silva (2006), the commercial interest in soya is related to its high protein and oil content, on average 40% and 20% respectively, and also to its productivity. Another important characteristic of soya, according to Amorim (2011), is the crop's ability to fix nitrogen symbiotically in its roots, thus eliminating the need for nitrogen fertilisers.

In the state of Tocantins, soya is the third largest crop in terms of gross production value. It is grown in the off-season (May-June), in irrigated floodplain conditions, under sub-irrigation (high water table), mainly in Formoso do Araguaia, and in the harvest period (November-December), in upland conditions (PELUZIO et al., 2005).

According to Costa et aL, (2004) genetic improvement programmes for the crop are essential to meet the growing demand for higher yields, making it possible to increase variability and, consequently, broaden the genetic base and select the best genotypes from a population capable of exceeding grain yield levels.

The success of a breeding programme depends on the existence of variability in the working population. In this sense, breeders have recommended intercrossing superior and divergent cultivars to form the base population, resulting in hybrid combinations with greater heterozygosity, so that in the segregating generations there is a greater chance of obtaining superior genotypes (CRUZ, 2007).

The use of multivariate techniques is one of the factors driving the increase in studies on genetic divergence between germplasm bank accessions. Multivariate analyses are based on algorithms or distance measures that simultaneously take into account numerous characteristics considered in germplasm characterisation and evaluation experiments (SUDRÉ et al., 2007; VILELA et al., 2008; ROCHA, 2010).

The purpose of clustering methods is to separate an original group of observations into various subgroups in order to achieve homogeneity within and heterogeneity between the subgroups. This methodology depends on the calculation of dissimilarity measures derived from quantitative and qualitative variables (CROSSA & FRANCO, 2004; ROCHA, 2010), and hierarchical and optimisation methods are widely used by plant breeders (CRUZ, 2007).

In optimisation methods, groups are established by perfecting a certain grouping criterion, differing from hierarchical methods in that the groups formed are mutually exclusive (CRUZ & REGAZZI, 2004). The optimisation method proposed by TOCHER adopts the criterion of keeping the average intra-group distance always lower than any inter-group distance (RAO, 1952).

The use of multivariate techniques to estimate the genetic divergence of the soya bean crop has been used in several studies (PELUZIO et al., 2009; SANTOS et al., 2011; ALMEIDA, PELUZIO & AFFÉRRI, 2011; SANTOS et al., 2013).

Despite numerous studies on genetic divergence in soya cultivation, there is no information on multivariate analysis in different parts of the plant, using agronomic, morphological and chemical characters, and their implications for genetic improvement programmes, which is what this work proposes.

MATERIALS AND METHODS

Two experiments were carried out in the experimental area of the Federal University of Tocantins (UFT) in the municipality of Palmas - TO, located at latitude 10^{s}45' S, longitude 47^{e}14' W at an altitude of 220m. The climate of the region, according to the Kõppen classification, is of the Aw type, tropical with a dry season. The first experiment was installed in the 2011/12 agricultural year (December 2011 to May 2012) and the second in the 2012/13 agricultural year (December 2012 to May 2013).

The chemical characteristics of the soil, classified as Latossolo Vermelho Amarelo Distrófico, according to the Brazilian Soil Classification System (SiBCS) (EMBRAPA, 2006), at a depth of 0-20 cm, in the two agricultural years are shown in Table II1-1.

Table II1-1. Soil analysis results from the Palmas experimental field in the 2011/12 and 2012/13 crop years.

Harvest years	pH	K (mg dm^{-3})	P* (mg dm^{-3})	M.O.	CTC	S. B (%)	Clay (%)	Silt (%)	Sand (%)	Classification
2011/12	4,9	42	25,5	1,8	3,08	79,80	20	8	72	Sandy loam
2012/13	5,8	95	18,3	1,4	5,54	71,12	23	6	71	Sandy loam

* Melich Method; O.M.: Organic Matter, CEC: Cation Exchange Capacity, B.S.: Base Saturation.

The experimental design used was randomised blocks, with 90 treatments and three replications. The treatments were laid out in a sub-subdivided plot design, with the agricultural years (2011/12 and 2012/13) allocated to the plots and 15 soya cultivars allocated to the sub-plots (BRS Valiosa, P98Y51, P98Y70, P99R03, M8527RR, M8925RR, M9144RR, M8867RR, TMG115RR, Conquista, Tucunaré, Pintado, M8866, M9350 and M9350RR).

TMG803), nine of which were transgenic and six conventional, and in the subplots the position of the pods on the main stem of the plant (lower third, middle third and upper third). These cultivars were chosen because they are widely used in the soil and climate conditions of the central region of the state of Tocantins.

The experimental plot consisted of four rows of 5.0m long soya plants, spaced 0.45m apart. When harvesting, only plants from the two centre rows were used, discarding 0.50m from the end of each row.

The soil was prepared by ploughing, harrowing and furrowing in both agricultural years. Sowing fertiliser, carried out after soil analysis and according to the crop's requirements, was 400 $kg.ha^{-1}$ of the NPK 05-25-15 formulation, followed by top dressing at stage R2 (full flowering) with 100 $kg.ha^{-1}$ of K2O, using potassium chloride as the source.

The sowing density was set at 14 plants per linear metre and thinning was carried out when necessary at 10 days after emergence. At the time of sowing, the beans were treated with fungicides (Carboxin + Thiram 200 SC - 2.5mL + 2.5mL of $water.kg^{-1}$), followed by inoculation of the beans with *Bradyrhizobium japonicum* strains, using the product Biomax® Premium Turfa - Soya (Strains 5079 and 5080), at a dosage of 60 g.50 kg^{-1} of soya beans.

Pest, disease and weed control was carried out as necessary. For pest control, the product Methamidophos (300 $g.ha^{-1)}$ was used at the R2, R4 and Re stages. To control Asian rust, two applications of the fungicide Tebuconazole 200 EC (0.5 $L.ha^{-1)}$ were made, one at R2 and the other at R4. Weeds were controlled with one application of Roundup Ready, at stages Ve and R2 of the soya bean, at a dose of 1.44 $L.ha^{-1}$.

In each experimental plot, 30 plants of each cultivar were harvested from the two centre rows, discarding 0.50 m from the ends. They were divided into the upper, middle and lower thirds, as shown in Figure 111-1 .

The pods of each variety and position were then sorted and packed in paper bags, which were labelled with the name of the cultivar and the position on the plant.

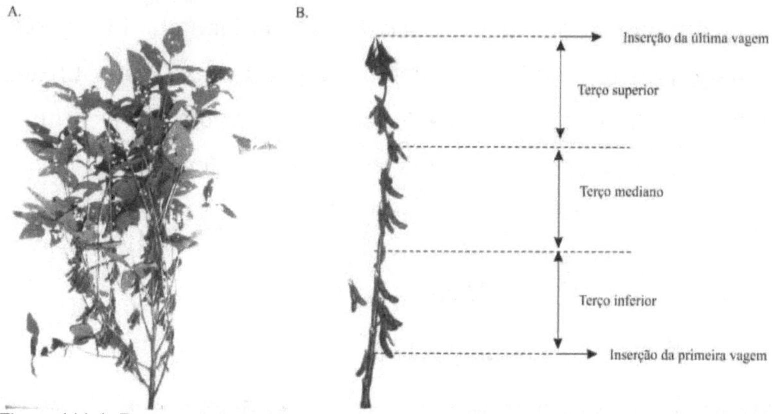

Figure 111-1. Representation of the soya plant harvested in the experimental field (A) and the section of the plant into thirds (lower, middle and upper) from the insertion of the first and last pod of the plant (B).

The agronomic characteristics of the field were determined after harvesting, on thirty plants randomised in the experimental plot, and the following evaluations were carried out on each part of the plant:

- Number of pods (NV): The number of pods produced by the thirty plants in each third was counted;
- Number of seeds (NS) per pod: The number of seeds produced in each part of the thirty plants was counted and the result divided by the number of pods (NV).
- Average unit weight of the grains (P): The grains from each part of the thirty plants were weighed on a SHIMADZU BL-3200H balance in order to weigh them.

  each third and then divided by the total number of seeds in each part (NS);

- Weight of 100 grains (WG): The average unit weight of the seeds in each part was multiplied by 100.

- Grain production per plant (PRO): Production was determined by the weight of all the seeds in each part.

- Diameter and length of grains: The diameter and length of 60 grains were measured randomly from the thirty plants that made up the experimental plot of each cultivar in each third of the plant, using a DIGIMESS caliper, according to the instructions in figure III-2.

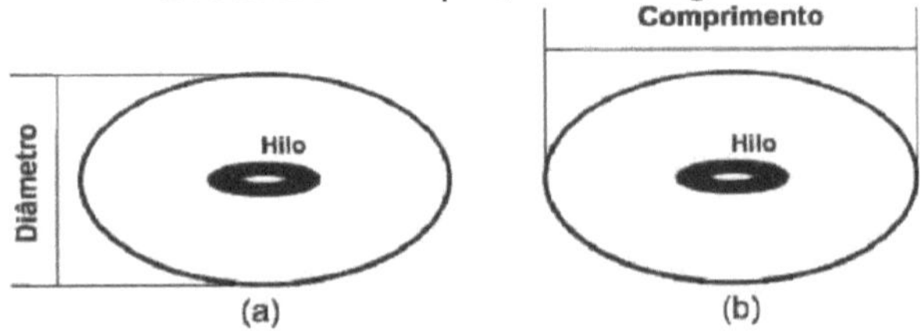

Figure III-2. Orientation of the soya beans during the tests to measure the smallest axis (a), diameter, and largest axis (b), length.

After evaluating the agronomic characteristics, the grains were ground and physico-chemical analyses were carried out.

The moisture content of the ground soya beans was determined on a GEHAKA infrared balance, model IVT440, at a temperature of 125ºC, for 2 min at 1.0000 ± 0.0200 in triplicate, according to (OLIVEIRA et al.,2010).

The dry protein content was determined using the Kjedhal method, converting the % of N into % of protein using a factor equal to 6.25 for cereals and oilseeds, according to the Adolfo Lutz Institute (IAL, 2005).

The dry lipid content for each cultivar in each third was determined using the Soxhlet method (IAL, 2005).

The protein and oil content data were subjected to the Kolmogorov-Smirnov normality test using the ASSISTAT statistical programme version 7.6 beta (SILVA & AZEVEDO, 2002).

Genetic divergence was analysed between the cultivars for each third of the plant (lower, middle and upper third) and also the divergence considering all the thirds of the plant together (the whole plant). The dissimilarity measures were determined according to the multivariate analysis model, which made it possible to obtain the dissimilarity matrix, the residual covariance matrix and the cultivars' means.

The multivariate procedure used was the TOCHER clustering method (Rao, 1952), using the generalised Mahalanobis distance (D^2) as a measure of dissimilarity. The criterion of (SINGH, 1981) was also used to quantify the relative contribution of these characteristics to genetic divergence.

Spearmann's correlation was carried out between the thirds of the plant and the whole plant, using the smallest Mahalanobis distances obtained for each parent in each third and in the whole plant, and the t-test was used to check the level of significance.

Statistical analyses were carried out using the Genes computer programme, version 2007 (CRUZ, 2007).

RESULTS AND DISCUSSIONS

The grouping of the cultivars by the TOCHER method in the whole plant and in each third of the plant (basal, middle and apical), obtained from the Mahalanobis distances, is shown in Table III-2. The whole-plant cluster analysis separated the cultivars studied into seven distinct groups.

Table 111-2. Grouping revealed by the TOCHER method, based on the generalised Mahalanobis distance of 15 soya cultivars, considering nine characteristics in two agricultural years 2011/12 and 2012/13 of the grains on the main stem.

Groups	Every plant	Lower third	Middle Third	Upper Third
1	4; 9; 2; 10; 15; 3	1; 13; 8; 14	4; 9; 15; 10; 2; 3; 5	3; 6; 7; 5; 15; 10; 4;
II	1; 13;8	2; 4; 9; 15; 7; 3; 5	1; 13; 8; 11	8; 13; 1
III	6; 7	10; 12; 6	7; 14	2; 9
IV	14	11	6	11
V	5		12	12
VI	12			
VII	11			

1-BRS Valiosa; 2-TMG803; 3-M8866; 4-P98Y51; 5-M8925RR; 6-M8867RR; 7-P98Y70; 8- TMG115RR; 9-P99R03; 10-M9144RR; 11-Pintado; 12-M8527RR; 13-Conquista; 14-Tucunaré and 15- M9350.

Group I included 06 genetically similar cultivars (40 per cent of all cultivars), indicating that the possible crossing of these cultivars with each other reduces the possibility of obtaining superior genotypes.

The cultivars Tucunaré, M8925RR, M8527RR and Pintado were isolated in groups IV, V, VI and VII, respectively. The formation of these groups is of fundamental importance when choosing parents, as the new hybrid combinations to be established must be based on the magnitude of their dissimilarities and the potential *per se of* the parents (CRUZ et aL, 2004). Cultivars in more distant groups are dissimilar and can be considered promising for artificial crosses.

The grouping revealed by the TOCHER method, based on the generalised Mahalanobis distance, for the lower third of the plant (table III-2) showed the formation of four distinct groups, with groups I and II represented by four and seven cultivars respectively. The cultivar PINTADO, as with the divergence in the whole plant (table III-2), was isolated in a single group.

In the cluster analysis in the middle third (table II1-2), five distinct groups were formed. The cultivars M8867RR and M8527RR were isolated in groups VI and V, respectively.

When the grouping of the cultivars in the upper third of the plant was assessed, the fifteen cultivars were separated into five distinct groups, which are shown in Table 111-2. In this third, group I was made up of seven genetically similar cultivars (46.67% of all cultivars). The cultivars Pintado and M8527RR were isolated in groups IV and V, respectively. Cultivar M8527RR was also isolated in the middle third (table 111-2).

For the middle third, the cultivars were separated into five distinct groups (table 111-2). Group I was made up of seven genetically similar cultivars (46.67% of all cultivars), indicating that the possible crossing of these cultivars with each other reduces the possibility of obtaining superior genotypes. The cultivars M8867RR and M8527RR were isolated in groups VI and V, respectively.

Cruz et al. (2004) suggest not involving individuals with the same dissimilarity pattern in crosses, so as not to restrict genetic variability and thus avoid negative effects on the gains to be obtained from selection. Establishing groups with genotypes that are homogeneous within and heterogeneous between groups is the starting point for a more detailed evaluation of these genotypes in order to make the most of them in breeding programmes (VIEIRA et aL, 2005).

Relative contribution

The relative contribution of each characteristic to genetic dissimilarity, according to Singh's method (1981) when analysing the whole plant and the three thirds, is shown in Table III-3.

For the whole plant, grain length contributed 53.5 per cent of the genetic divergence, followed by oil content (17.3 per cent) and 100-seed weight (8.7 per cent).

Table III-3. Relative contribution of characters to the genetic dissimilarity of fifteen soya cultivars in two agricultural years 2011/12 and 2012/2013, for the whole plant, using the method proposed by SINGH (1981).

Variables analysed	Amount of contribution in %			
	Lower third	Middle third	Upper third	The whole plant
Number of pods per part	1,39	7,83	8,39	7,7
Number of seeds per pod	1,37	2,97	2,36	1,1
Unit weight	4,52	3,78	5,20	3,6
Weight of 100 seeds	14,26	5,51	13,55	8,7
Production	5,97	1,36	1,26	0,5
Grain length	29,01	50,44	40,46	53,5
Grain diameter	13,91	4,23	7,94	3,7
Protein Percentage	7,01	3,97	8,55	3,9
Percentage of oil	22,56	19,91	12,29	17,3

The relative contribution of each characteristic to the genetic dissimilarity for the lower third (Table III-5) showed that the characteristics that contributed most to the genetic divergence of the cultivars for this third was grain length (29.01%), followed by oil percentage (22.56%). These results were similar to the relative contribution obtained for the whole plant (Table III-3).

In the middle third, the relative contribution of each trait to genetic dissimilarity was similar to that found for the whole plant and the lower third (Table III-3), showing a contribution of 50.44% (grain length) and 19.91% (oil content). The upper third of the plant, as well as the other thirds of the plant and the whole plant, showed the greatest contributions to the grain length

(40.46%) and oil content (12.29%) characteristics.

Spearmann correlation

The estimates of Spearmann's correlation coefficients using the smallest Mahalanobis distances obtained for each cultivar, between the thirds and between the thirds and the whole plant, are shown in Table II1-4.

Table 111-4. Estimates of the Spearmann correlation coefficients of fifteen soya cultivars, using the smallest Mahalanobis distances obtained for each cultivar, between the thirds of the plant and between the thirds and the whole plant, in the 2011/12 and 2012/2013 harvests, in Palmas.

Plant parts	Lower Third	Middle third	Upper third	Whole plant
Lower third		0,51*	-0,07	0,74*
Middle third			0,04	0,85*
Upper third				0,11

* Significant at 5% by t-test.

Positive and significant correlations were observed between the lower and middle thirds (r = 0.51) and between the lower and middle thirds with the whole plant (r = 0.74 and r = 0.85, respectively) revealing, for these combinations, that the characteristics originating from different parts of the plant, expressed from the smallest Mahalanobis distances, did not result in changes in the relative position of the cultivars. Thus, for these pairs, similarity would be expected in terms of composition and the number of groups to be formed by the TOCHER method, which would make it possible to use the lower and middle thirds instead of the whole plant in multivariate procedures to study genetic diversity for selection purposes in breeding programmes.

However, when TOCHER groupings were carried out for each of these thirds, compared to the whole plant, there was no agreement as to the composition and number of groups formed, indicating that the use of the correlation coefficient is inefficient for this purpose

Spearmann's ratio, obtained from the smallest Mahalanobis distances obtained for each parent.

CONCLUSIONS

Grain length and grain oil content, regardless of the part of the plant studied, are the characteristics with the greatest contributions to the dissimilarity of the cultivars studied;

The composition and number of groups formed using the TOCHER method did not agree between the thirds, nor did they agree with the analysis of the whole plant;

In soybean breeding programmes aimed at selecting superior genotypes, given the discrepancy between the results obtained in the thirds of the plant, as well as their low association with the analysis of the whole plant, it is preferable to carry out a multivariate analysis of the whole plant for the characters.

BIBLIOGRAPHICAL REFERENCES

ALMEIDA, R.D.; PELUZIO, J.M.; AFFÉRRI, F.S. Genetic divergence between soya cultivars, under irrigated floodplain conditions, in southern Tocantins State. **Revista Ciência Agronómica,** v.42, n.01, p. 108-115, 2011.

AMORIM, F.A. **Effect of the CP4 EPSPS gene on oil productivity in soya populations.** 2011. Thesis (PhD in Genetics and Plant Breeding) - Luiz de Queiroz College of Agriculture, University of São Paulo, Piracicaba, 2011. Available at:< http://www.teses.usp.br/teses/disponiveisZ11/11137/tde-13092011 - 083447/>. Accessed on: 03 Mar 2013.

BONATO, E.R.; BONATO, A.L.V. **Soya in Brazil: History and statistics.** Londrina: EMBRAPA-CNPSo, 1987. 61 p.

BONATO, E.R.; BERTAGNOLLI, P.F.; LANGE, C.E.; RUBIN, S. de A.L.Oil and protein content in soya genotypes developed after 1990. **Pesquisa Agropecuária Brasileira,** v. 35, p. 2391-2398, 2000.

BONETT, L.P.; GONÇALVES-VIDIGAL, M.C.; SCHUELTER, A.R.; VIDIGAL FILHO, P.S.; GONELA, A.; LACANALLO, G.F. Genetic divergence in common bean germplasm collected in the State of Paraná, Brazil. **Semina: Ciências Agrárias,** v.27, p.547-560, 2006.

BURRIS, J.S.; EDJE, O.T.; WAHAB, A.H. Effect of size on seedling performance in soybeans: Seedling growth and photosynthesis and field performance. **Crop Science,** v.13, n.2, p.207-210, 1973.

CEOLIN, A.C.G.; VIDIGAL, M.C.G.; VIDIGAL FILHO, P.S.; KVITSCHAL, M.V.; GONELA, A.; SCAPIM, C.A. Genetic divergence of the common bean (Phaseolus vulgaris L.) group carioca using morpho-agronomic traits by multivariate analyses. **Hereditas,** v. 144, p 1-9, 2007.

COSTA, M.M.; MAURO, A.O.D.; UNÊDA-TREVISOLI, S.H.; ARRIEL, N.H.C.; BÁRBARO, LM.; MUNIZ, F.R.S. Genetic gain by different selection criteria in soybean segregating populations. Pesquisa Agropecuária Brasileira, Brasília, v.39, n.11, p.1095-1102.

CLEMENTE, T.E.; CAHOON, E.B. Soybean Oil: Genetic Approaches for Modification of Functionality and Total Content. **Plant Physiology,** v. 151, p. 1030-1040, 2009.

GROSSA, J.; FRANCO, J. Statistical methods for classifying genotypes. **Euphytica,** v. 137, n. 1, p. 19-37, 2004.

CRUZ, C.D.; REGAZZI, A.J.; CARNEIRO, P.C.S. **Biometric models applied to genetic improvement.** 3.ed. Viçosa: UFV. 279p. 2004.

CRUZ, C.D. **Programa Genes: Aplicativo computacional em genética e estatística.** Windows version - 2007, Viçosa, UFV.

CRUZ, C.D.; REGAZZI, A.J. **Biometric models applied to genetic improvement.** Viçosa, MG: University Press, 2004, 480 p.

EDMAR, V.S.; REIS, M.S.; SEDIYAMA, T. Non-parametric analysis of seed health and elimination indices and classification of soya genotypes. **Pesquisa Agropecuária Brasileira,** v. 43, p. 341-348, 2008.

EMBRAPA - Brazilian Agricultural Research Corporation. National Soil Research Centre. **Brazilian soil classification system.** Rio de Janeiro, 2006. 306p.

GOMES, P. **A soja.** 5th edition. 2nd reprint. São Paulo: Nobel, 1988. p.152.

GONTIA, A.S.; AWASTHI, M.K. Effect of seed grading by size on various seed vigour attributes, morphophysiological characters and seed yield in soybean *[Glycine max* (L.) Merrill] genotypes. **Seed Research,** v.27, n.1, p.333-350, 1999.

ADOLFO LUTZ INSTITUTE. **Analytical standards:** chemical and physical methods for food analysis. 4. ed. São Paulo, 2005.1018 p.

KOLAK, I.; HENNEBERG, R.; MILAS, S.; RADOSEVIC, J.; SATOVIC, Z. Influence of seed on yield and yield components in different cultivars. **Polioprivredna,** v.57, p.519-526,1992.

KRZYZANOWSKI, F.C.; FRANÇA-NETO, J. B.; COSTA, N.P. da; HENNING, A.A; VIEIRA, B.G.T.L. Influence of seed size on soya crop productivity. In:

SOYBEAN RESEARCH MEETING OF THE CENTRAL REGION OF BRAZIL, 26, 2004, Ribeirão Preto. **Proceedings...** Londrina: Embrapa Soja, 2004, p.318-319 (Embrapa Soja. Documentos, 234).

KRZYZANOWSKI, F.C.; FRANÇA-NETO, J. COSTA, N.P. da; HENNING, A.A.; VIEIRA, B.G.T.L Influence of seed size on soya crop productivity. In: REUNIÃO DE PESQUISA DE SOJA DA REGIÃO CENTRAL DO BRASIL, 27., 2005, Cornélio Procópio. **Proceedings...** Londrina: Embrapa Soja, 2004, p.567-568 (Embrapa Soja. Documentos, 257).

KRZYZANOWSKI, F.C.; FRANÇA-NETO, J. B.; FARIAS, J.R.B.; COSTA, N.P.; HENNING, A.A; Influence of seed size on soya bean yield. In: REUNIÃO DE PESQUISA DE SOJA DA REGIÃO CENTRAL DO BRASIL, 30., 2008, Cornélio Procópio. **Proceedings...** Londrina: Embrapa Soja, 2004, p.337-338. (Embrapa Soja. Documentos, 304).

LIMA, A.M.M.P.; CARMONA, R. Influence of seed size on the productive performance of soya beans. **Revista Brasileira de Sementes,** v. 21, n.1, p. 157-163, 1999.

OLIVEIRA, M.A.; CARRÃO-PANIZZI, M.C.; MANDARINO, J.M.G.; LEITE, R.S.; CAMPOS FILHO, P.J.; VICENTINI, M.B. Quantification of reducing sugars, oligosaccharides and starch in special genotypes/cultivars of soya *beans (Glicine Max* (L) Merril) used for human consumption. **Brazilian Journal of Food Technology,** v. 13, n. 01, p. 23-29, 2010.

PELUZIO, J.M.; ALMEIDA JUNUOR, D.; FRANCISCO, E.R.; FIDELIS, R.R.; RICHTER, L.H.M.; RICHTER, C.A.M.; BARBOSA, V.S. Behaviour of soya cultivars in the south of the state of Tocantins. **Bioscience Journal.** Uberlândia, MG. v.1, n.3, p. 113-117, 2005.

PELUZIO, J.M.; VAZ-DE-MELO, A.; AFFÉRRI, F.S.; SILVA, R.R.; BARROS, H.B.; NASCIMENTO, LR.; FIDELIS, R.R. Genetic variability among soya cultivars under different soil and climate conditions. **Pesquisa Aplicada & Agrotecnologia** v.2 n.3, p. 21-29, 2009.

RAO, C.R. **Advanced statistical methods in biometric research.** New York: John Willey, 1952, 390 p.

ROCHA, M.C; GONÇALVES, L.S.A.; RODRIGUES, R.; SILVA, P.R.A.; CARMO, M.G.F.; ABBOUD, A.C.S. Use of the Gower algorithm to determine genetic divergence between cherry tomato accessions. **Acta Scientiarum. Agronomy,** v. 32, n. 3, p. 423-431,2010.

SANTOS, E.R.S.; BARROS, H.B.; FERRAZ, E.C.; CELLA, A.J.S.; CAPONE, A.; SANTOS, A.F.; FIDELIS, R.R. Divergence between soya genotypes grown in irrigated floodplains. **Revista Ceres,** v.58, n.6, 2011.

SANTOS, E.R.; SANTOS, A.F.; CAPONE, A.; SANTOS, W.R.; MOURA, S.G.; BARROS, H.B. Genetic dissimilarity between soybean genotypes cultivated in lowland irrigated inter-cropping. **Journal of Biotechnology and Biodiversity,** v.4, n.3: p.222-231,2013.

SCAPIM, C.A.; PIRES, I.E.; CRUZ, C.D.; AMARAL JÚNIOR, A.T.; BRACCINI, A.L.; OLIVEIRA, V.R. Evaluation of genetic diversity in Eucalyptus camaldulensis Dehnh, by means of multivariate analysis. **Revista Ceres,** v.6, p.347-356, 1999.

SEDIYAMA, T.; TEIXEIRA, R.C.; REIS, M.S. **Improving soya beans.** In: BORÉM, A. Improvement of cultivated species. Viçosa: UFV, 2005. 969 p.

SINGH, D. The relative importance of characters affecting genetic divergence. **Indian Journal Genetics Plant Breeding,** v.41, p.237-245, 1981.

SOUZA, E.L. de. **Quality of soya beans marketed by the COPAGRIL agro-industrial cooperative in Paraná.** 2006. 34p. Dissertation (Master's Degree in Seed Science and Technology) - Federal University of Pelotas. Pelotas.

SUDRÉ, C.P.; GONÇALVES, L.S.A.; RODRIGUES, R.; AMARAL JÚNIOR, A.T.; RIVA-SOUZA, E.M.; BENTO, C.S. Genetic variability in domesticated Capsicum spp. as assessed by morphological and agronomic data in mixed statistical analysis. **Genetics and Molecular Research,** v. 9, n. 1, p. 283- 294, 2007.

VELLO, N.A.; SILVA, L.A.S. Genetics seeks to meet growing human consumption. **Visão Agrícola,** year 3, n.5, p.60 - 62, 2006.

VIANA, C.F.A.; SILVA, M.A.; PIRES, A.V.; LOPES, P.S.; PIASSI, M. Estudo da divergência genética entre quatro linhagens **de** matrizes **de frango de** corte utilizando técnicas **de** análise multivariada**, Revista Brasileira de Zootecnia,** v.29, n.4, p.1074-1081,2000.

VIEIRA, E.A.; FIALHO, J.F.; FALEIRO, F.G.; FUKUDA, W.M.G.; JUNQUEIRA, N.T.V. Genetic variability for morphological characters among accessions of the Embrapa Cerrados cassava germplasm bank. In: XI Brazilian Cassava Congress, Campo Grande. **Proceedings...,** CBM, CD ROM. 2005.

VILELA, F.O.; AMARAL JÚNIOR, A.T.; PEREIRA, M.G.; SCAPIM, C.A.; VIANA, A.P.; FREITAS JÚNIOR, S.P. Effect of recurrent selection on the genetic variability of the UNB-2U popcorn population using RAPD markers. **Acta Scientiarum. Agronomy,** v. 30, n. 1, p. 25-30, 2008.

ZANON, G.D. **Protein and Oil Content in Soya Beans Under Different Types of Management.** 2007. 62 f. Thesis (Doctorate) - Universidade Federal da Grandbe Dourados, Dourados, MS, 2007.

FSC
www.fsc.org
MIX
Papier | Fördert gute Waldnutzung
FSC® C083411

Kolibri 360 GmbH
Ferdinand-Jühlke-Straße 7
99095 Erfurt, Deutschland
produktsicherheit@kolibri360.de

Druck:
CPI Druckdienstleistungen GmbH
im Auftrag der
Kolibri 360 GmbH
Ferdinand-Jühlke-Str. 7
99095 Erfurt